BEIHEFTE ZUR
NOVA HEDWIGIA

HEFT 41

CATALOGUE OF THE LICHENS OF AUSTRALIA

EXCLUSIVE OF TASMANIA

BY

WILLIAM A. WEBER
AND
CLIFFORD M. WETMORE

3301 LEHRE

VERLAG VON J. CRAMER

1972

Printed in Germany 31.V.1972
by Strauss & Cramer GmbH, 6901 Leutershausen
ISBN 3 7682 5441 0

ACKNOWLEDGEMENTS

The senior author wishes to express his gratitude to the University of Colorado Council on Research and Creative Work for its award of a Faculty Fellowship during the year 1967-68, the Research School of Pacific Studies of the Australian National University, Canberra, for providing him with its Research Fellowship and field support during that time, and to Dr. Donald McVean, in whose company the field studies were performed and who provided most valuable insight into the ecological behavior of the Australian lichens. We are indebted to the staffs of the Advanced Studies Library of the Australian National University, the Royal Botanic Gardens, Sydney and the C.S.I.R.O. Library, Canberra, for invaluable bibliographic assistance. A significant portion of the project was effected through the normal curatorial procedures of the University of Colorado Museum and its assistants. We are indebted also to the University of Colorado College of Arts and Sciences for the preparation of the final manuscript for publication.

TABLE OF CONTENTS

PREFACE

The senior author began to prepare a comprehensive catalogue of the Australian lichens in 1966 in anticipation of the opportunity afforded for field work there by a year's tenure as Research Fellow at the Australian National University at Canberra. An attempt was made to record systematically all of the records of lichens scattered in a vast taxonomic literature. The junior author already had published a catalogue of the lichens of Tasmania (Wetmore 1963) and in so doing accumulated a large number of additional bibliographic citations pertaining to Australia. We agreed to cooperate fully in the venture, the results of which are presented here.

No catalogue of the Australian lichens has ever been published before, but several works by Australian authors provided invaluable primary sources of historical and bibliographic material. These are Cheel's Bibliography of Australian Lichens (1903) which attempted to list all of the major publications dealing with these plants, and Maiden's (1908-12) and Wilson's (1890) papers on the history and personalities of Australian lichenology. We have used Cheel as our starting point of investigation. All of those papers mentioned by Cheel are so identified by his number following the entry in our bibliography. The history of lichenological work in Australia, however, is a major topic in itself and lies beyond the scope of this paper.

The purpose of the catalogue is primarily bibliographic rather than taxonomic. It is presented with the aim of aiding Australian lichenologists in discovering what taxa have been reported and where the reports may be found in the literature. It does not pass judgment on the validity of reports; indeed, the reports of a great many European species reported for Australia were based on patently incorrect identifications. Only a few new records are presented here for the first time. These are chiefly alpine species discovered by the senior author in the Snowy Mountains in 1967-68.

In order to reduce the bulk of the work and to present some semblance of digestion of the raw material, some subjective decisions have been introduced. Reliance generally has been placed on the synonymic interpretations of Zahlbruckner's Catalogus except where we have real evidence that he was incorrect. At the same time, we have made it possible for the user to recapture taxonomic synonyms otherwise lost in the list by providing a separate index to them.

The arrangement of the catalogue is alphabetical by genus and species. Arrangement by families was contemplated but discarded because of so much uncertainty at the present time concerning generic relationships. Taxa that were described from Australian stations are indicated by the asterisk (*) preceding the name. The main entry is followed by its bibliographic citation. Following the main entry are synonyms under which reports have been made. The province of the type locality is given for those taxa with Australian types. The authors and dates of those papers in which the taxa are mentioned follow the nomenclature, and these are followed by abbreviations of the provinces of Australia from which the taxa have been reported. Records that are presented here for the first time are presented with full citation of the specimen data and the international herbarium abbreviation. The author and date citations refer in turn to the bibliography provided later on in the paper. In the bibliography all papers that are not identified by an asterisk have been seen by the authors.

In certain generic pairs, such as *Sticta* and *Pseudocyphellaria*, *Blastenia* and *Caloplaca*, some taxa cannot be listed in the proper genus under their original and often current basionyms. Rather than to propose new combinations for lichens the types of which we have not examined we suggest the proper taxonomy by their position in the list, enclosing the original generic name in parentheses.

CATALOGUE OF THE LICHENS OF AUSTRALIA

ACAROSPORA

A. CERVINA (Ach.) Mass., Ricerch. Auton. Lich. p. 28. 1852. Lecanora cervina Ach., Syn. Lich. p. 188. 1814. Placodium cervinum var. percaenum Muell.-Arg., Hedwigia 31: 194. 1892. Magnusson (1929). WA.

*A. FERDINANDII Hue, Nouv. Arch. Mus. Paris (5) 1: 160. 1909. Placodium ferdinandii Muell.-Arg., Flora 64: 508. 1881. "Central Australia near Eureo." Magnusson (1929); F. Mueller (1881); Muell.-Arg. (1881b). CA, WA.

*A. FUSCORUFA F. Wils. ex H. Magn., Kongl. Svenska Vetensk.-Akad. Handl. (3) 7 (1): 309. 1929. V (Dunkeld). Isotype seen (MEL) = A. fuscata (Schrad.) Arn!

*A. MACROCARPA H. Magn., Kongl. Goetob. Vetensk. Vitterh.-Samh. Handl. F. 6, ser. B, VI, no. 17: 25. 1956. "Australia, terricola."

*A. NEGLIGENS H. MAGN., Kongl. Svenska Vetensk.-Akad. Handl. (3) 7 (4): 58. 1929. WA.

A. REAGENS A. Zahlbr., Beih. Bot. Centralbl. 13: 162. 1902. South Australia: Nullarbor Plain, McVean 6637 (COLO).

A. SCHLEICHERI (Ach.) Mass., Ricerch. Auton. Lich. p. 27. 1852. Urceolaria schleicheri Ach., Lichenogr. Univ. p. 332. 1810. U. citrina Tayl., London J. Bot. 6: 158. 1847. Acarospora citrina A. Zahlbr. ex Rech., Denkschr. K. Akad. Wiss. Math. Naturwiss. Kl. 88: 28. 1911. A. initialis H. Magn., Kongl. Svenska Vetensk.-Akad. Handl. (2) 7 (1): 48 1929. A. novae-hollandiae H. Magn., l.c. p. 89. A. oculata H. Magn., l.c. p. 51. A. Wilsonii H. Magn., l.c. p. 71. Magnusson (1929); Muell.-Arg. (1888d, 1892b, 1893a); Weber (1968). V. WA.

ALECTORIA

*A. AUSTRALIENSIS C. Knight ex Bailey, Proc. Roy. Soc. Queensland 1: 90. 1884. Q. Du Rietz (1926); Muell.-Arg. (1891a); Shirley (1889a, d, 1893a. According to Muell.-Arg. (1891), this is not a lichen. Shirley refers to it as the "vegetable horse-hair," which is Nostoc flagelliforme or a similar alga.

A. NIGRICANS (Ach.) Nyl., Lich. Scandin. p. 71. 1861. Cornicularia ochroleuca beta C. nigricans Ach., Lichenogr. Univ. p. 615. 1810. NSW: summit of Mt. Stillwell, Kangaroo Range, Mt. Kosciusko State Park, 6000 ft., Dec. 1966, McVean L-47262 (COLO).

A. PUBESCENS (L.) Howe f., Classific. Famil. Usneae, p. 23. tab. IX, fig. 2. 1912. Lichen pubescens L., Sp. Pl. p. 1155. 1753. NSW: Mount Kosciusko, trail to Northcote Pass, 7000 ft., 10 Nov. 1967, Weber & McVean L-47192 (COLO).

ANAPTYCHIA

A. BARBIFERA (Nyl.) Trev., Flora 44: 52. 1861. Physcia barbifera Nyl., Syn. Lich. 1: 416. 1860. Bailey (1881); F. Mueller (1881); Shirley (1889a). Q. Kurokawa (1962) does not list for Australia.

A. COMOSA (Eschw.) Mass., Mem. Lichenogr. p. 39. 1853. Parmelia comosa Eschw. ex Martius, Icones Plant. Cryptog. 2: 25. 1828-34. Physcia comosa Nyl., Mem. Soc. Imp. Sci. Nat. Cherbourg 3: 175. 1855. Shirley (1893a); Wilson (1891c). Q.

A. DIADEMATA (Tayl.) Kurok., Nova Hedwigia Beih. 6: 28. 1962. Parmelia diademata Tayl, London J. Bot. 6: 165. 1847. A. hypoleuca var. diademata A. Zahlbr., Ann. K. K. Naturhist. Mus. 11(2): 191. 1896. Physcia speciosa var. major (Nyl.) Muell.-Arg., Flora 70: 60. 1887. Krempelhuber (1880); Kurokawa (1962); F. Mueller (1881); Zahlbruckner (1896). NSW, V.

A. HYPOLEUCA (Muehl.) Mass., Atti Imp. Regia Istit. Veneto (3) 5: 249. 1860. Parmelia hypoleuca Muehl., Cat. Pl. Amer. Sept. p. 105. 1813. Physcia speciosa var. hypoleuca Nyl., Syn. Lich. 1: 417. 1860. Knight (1882); Krempelhuber (1880); Wilson (1887). NSW, V. Kurokawa (1962) does not list for Australia, suggesting that old reports can be attributed to misidentification.

A. LEUCOMELAENA (L.) Vain., Etud. Lich. Brésil 1: 128. 1890. Lichen leucomelos L., Sp. Pl. ed. 2, 2: 1613. 1763. Physcia leucomela Michx., Fl. Bor.-Am. 2: 327. 1803. Hellbom (1896); Krempelhuber (1880, "var. subcomosa Nyl."); F. Mueller (1881); Nylander (1857); Shirley (1892b, 1893a); Stirton (1899b). Q.

A. LEUCOMELAENA var. ANGUSTIFOLIA (Mey. & Flot.) Muell.-Arg., Bot. Jahrb. 20: 249. 1894. Parmelia leucomela var. angustifolia Mey. & Flot., Nova Actorum Acad. Caes. Leop.-Carol. Nat. Cur. 19 (Suppl.): 221. 1843. Knight (1884b); Shirley (1889a). Q.

A. LEUCOMELAENA var. MULTIFIDA Vain., Etud. Lich. Brésil 1; 128. 1890. Parmelia leucomela f. multifida Mey. & Flot., Nova Actorum Acad. Caes. Leop. Carol. Nat. Cur. 19 (Suppl.): 221. 1843. Muell.-Arg. (1891a). Q.

A. OBESA (Pers.) A. Zahlbr. var. CAESIOCROCATA (Nyl.) A. Zahlbr., Cat. Lich. Univ. 7: 734. 1931. Physicia obesa f. caesiocrocata Nyl., Syn. Lich. 1: 418. 1860. Bailey (1881); Shirley (1889a). A.

A. OBSCURATA (Nyl.) Vain., Acta. Soc. Faun. Fl. Fenn. 7: 137. 1890. Kurokawa (1962); Weber (1969). Q.

A. OPHIOGLOSSA (Tayl. in Hook. f.) Kurok. f. ALBOCILIATA (Nyl.) Kurok., J. Jap. Bot. 35: 357. 1960. Physcia leucomela f. albociliata Nyl., Ann. Sci. Nat., Bot. (4) 19: 309. 1863. Kurokawa (1960). Sine loc.

A. PODOCARPA (Bel.) Mass., Atti Imp. Regia Istit. Veneto (3) 5: 249. 1860. Parmelia podocarpa Bel., Voy. Ind.-Or. 2: 122 1846. Physcia podocarpa Nyl., Flora 52: 322. 1869. Krempelhuber (1880); F. Mueller (1881). Sine loc.

A. RAVENELII (Tuck.) A. Zahlbr., Cat. Lich. Univ. 7: 737. 1931. Physcia ravenelii Tuck., Syn. N. Am. Lich. 1: 68. 1882. Stirton (1889b as "P. crispa var. ravenelii"). Q.

A. SPECIOSA (Wulf.) Mass., Mem. Lichenogr. p. 36. 1853. Lichen speciosus Wulf. ex Jacq., Collect. Bot. 3: 119. 1789. Physcia speciosa Nyl, Act. Soc. Linn. Bordeaux 21: 307. 1856. Bailey (1881, 1883); Hellbom (1896); Hue (1890-92); F. Mueller (1881); Nylander (1857, 1886); Shirley (1889a, 1892b); Stirton (1899b); Tate (1882); Turner (1905); Willis (1953); Wilson (1890b). NSW, Q, SA, WA, V.

*A. SPECIOSA var. ANGUSTIFOLIA (Muell.-Arg.) A. Zahlbr., Cat. Lich. Univ. 7: 741. 1931. Physcia speciosa var. angustifolia Muell.-Arg., Flora 66: 78. 1883. Q.

A. SPECIOSA f. SOREDIIFERA (Muell.-Arg.) A. Zahlbr., Cat. Lich. Univ. 7: 741. 1931. Physcia speciosa f. sorediifera Muell.-Arg., Flora 67: 688. 1884. Muell.-Arg. (1887c, 1892a); Shirley (1893a); Willis (1953). Q. WA.

A. SPECIOSA f. SOREDIOSA (Muell.-Arg.) A. Zahlbr., Cat. Lich. Univ. 7: 741. 1931. Physcia speciosa f. sorediosa Muell.-Arg., Flora 66: 78. 1883. Q.

ANTHRACOTHECIUM

A. AMPHITROPUM Muell.-Arg., Flora 66: 246. 1883. Muell.-Arg. (1891a, 1895b); Shirley (1891a). Q.

A. AURANTIACUM (Eschw. ex Mart.) Muell.-Arg., Flora 67: 664. 1884. Verrucaria aurantiaca Eschw. ex Mart., Icon. Pl. Crypt. 2: 14. 1828-34. Muell.-Arg. (1895b); Shirley (1895). Q.

A. CONFINE (Nyl.) Muell.-Arg., Linnaea 63: 45. 1880. Verrucaria confinis Nyl., Ann. Sci. Nat., Bot. (4) 3: 174. 1885. Muell.-Arg. (1895b); Shirley (1893a). Q.

*A. DESQUAMANS Muell.-Arg., Flora 71: 48. 1888. Muell.-Arg. (1895b); Shirley (1890). Q.

A. DOLESCHALLII Mass. ex Kremp., Verh. Zool.-Bot. Ges. Wien 21: 863. 1871. Muell.-Arg. (1895b); Shirley (1893a). Q.

*A. LIBRICOLUM (Fee) Muell.-Arg., Linnaea 63: 43. 1880. Pyrenula libricola Fée, Suppl. Essai Crypt. Ecorc. Officin. p. 82. 1837. Bailey (1881, 1883); F. Mueller (1881); Muell.-Arg. (1895b). Q.

A. MACROSPORUM (Hepp) Muell.-Arg., Linnaea 63: 44. 1880. Verrucaria macrospora Hepp ex Zoll., System. Verzeichn. Ind. Arch. Gesamm. Pfl. p. 9. 1854. Muell.-Arg. (1895b). Q.

*A. MONOSPORUM Muell.-Arg. Bull. Herb. Boiss. 3: 327. 1895. V.

A. OCHROTROPUM (Nyl.) A. Zahlbr., Cat. Lich. Univ. 1: 465. 1922. Verrucaria denudata f. ochrotropa Nyl., Bull. Soc. Linn. Normand. (2) 2: 129. 1868. Anthracothecium denudatum var. ochrotropum Muell.-Arg. J. Bot. (Desvaux) 7: 111. 1893. Muell.-Arg. (1895b); Shirley (1895). Q.

*A. OCULATUM Muell.-Arg., Nuovo Giorn. Bot. Ital. 23: 404. 1891. Bailey (1891); Muell.-Arg. (1891a, 1895b); Shirley (1891b). Q.

*A. OLIGOSPORUM Muell.-Arg., Flora 71: 48. 1888. Muell.-Arg. (1895b); Shirley (1889b, as "oligospermum," 1890, 1895). Q.

A. PARVINUCLEUM (Mey. & Flot.) A. Zahlbr., Cat. Lich. Univ. 1: 466. 1922. Verrucaria parvinuclea Mey. & Flot., Nova Actorum Acad. Caes. Leop. Carol. Nat. Cur. 19: 231. 1843. A. denudatum (Nyl.). Muell.-Arg., Linnaea 63: 45. 1880. Muell.-Arg. (1895b); Shirley (1895). Q.

*A. PYRENULOIDES (Mont.) Muell.-Arg. Linnaea 63: 44. 1880. Trypethelium pyrenuloides Mont., Ann. Sci. Nat., Bot. (2) 19: 69. 1843. Muell.-Arg. (1891a, 1895b); Shirley (1891a). Q.

A. SINAPISPERMUM (Fée) Muell.-Arg., Linnaea 63: 45. 1880. Verrucaria sinapisperma Fée, Essai Crypt. Ecorc. Officin. p. 86. 1824. Bailey (1881, 1883); F. Mueller (1881); Muell.-Arg. (1895b); Shirley (1890). Q.

*A. SUBVARIOLOSUM Shirley, Lich. Fl. Queensland 4: 182. 1890. Verrucaria subvariolosa Knight ex Shirley, l.c. Q.

A. THWAITESII (Leight.) Muell.-Arg., Linnaea 63: 44. 1880. Verrucaria thwaitesii Leight., Trans. Linn. Soc. London 27: 184. 1869. Muell.-Arg. (1891a, 1895b); Shirley (1891a). Q.

(Anthracothecium)

A. VARIOLOSUM (Pers.) Muell.-Arg.., Linnaea 63: 44. 1880. Pyrenula variolosa Pers. in Gaudich., Voy. autour du monde, Bot. p. 181. 1826. Muell.-Arg. (1895b). Q.

ANZIA

A. ANGUSTATA (Pers. in Gaudich.) Muell.-Arg., Flora 72: 507. 1889. Parmelia angustata Pers. in Gaudich., Voy. Uranie, Bot. p. 195. 1826. P. inaequalis Tayl., London J. Bot. 6: 169. 1847. P. angustata var. moniliformis Stirt., Trans. Proc. New Zealand Inst. 32: 91. 1899. Crombie (1880); Hue (1890-92); Maiden (1898); F. Mueller (1881); Mueller.-Arg. (1889); Stirton (1899a); Weber (1969); Wilson (1887, 1889b). NSW, V. Zahlbruckner (Catalogus 7: 676) erred in citing Parmelia angustata Roehl. as "1813." The correct date is 1831, so P. angustata Pers. is not a later homonym.

*A. ANGUSTATA var. ISIDIELLA (Stirt.) A. Zahlbr., Cat. Lich. Univ. 6: 276. 1929. Parmelia angustata var. isidiella Stirt., Trans. Proc. New Zealand Inst. 32: 81. 1899. Sine loc.

*A. COLPODES (Ach.) Stizenb. var SPHAEROIDISPORA F. Wils., Victoria Naturalist 6: 69. 1889. Sine loc.

*A. WILSONII Raes., Ann. Bot. Soc. Zool.-Bot. Fenn. "Vanamo" 20 (3): 2. 1944. Sine loc. Weber (1971). NSW.

ARTHONIA

*A. ALBOFARINOSA Stirt. ex Bailey, Queensland Agr. J. 5: 488. 1899. Q.

*A. AMOENA Muell.-Arg., Bull. Herb. Boiss. 3: 322. 1895. Shirley (1896). Q.

*A. BANKSIAE Muell.-Arg., Bull. Herb. Boiss. 1: 59. 1893.

A. CINNABARINA (DC.) Wallr., Fl. Crypt. German. 3: 320. 1831. Arthonia gregaria Koerb., Syst. Lich. German. p. 291. 1855. Bailey (1881-1883); F. Mueller (1881); Shirley (1888b, 1889a); Wilson (1887). Q, V.

A. CINNABARINA var. COCCINEA (Floerke) A. Zahlbr., Cat. Lich. Univ. 2: 24. 1922. Conioloma coccineum Flk. in Mart., Fl. Cryptog. Erlangens. p. 284. 1817. Arthonia gregaria var. adspersa Muell.-Arg. Flora 71: 524. 1888. Bailey (1891a); Muell.-Arg. (1891b, 1893); Shirley (1891a). Q, V.

A. CINNABARINA var. NUDATA (Muell.-Arg.) A. Zahlbr., Cat. Lich. Univ. 2: 26. 1922. A. gregaria var. nudata Muell.-Arg., Hedwigia 30: 183. 1891. V.

A. CINNABARINA var. PURPUREA (Eschw.) A. Zahlbr., Cat. Lich. Univ. 2: 29. 1922. Conioloma coccineum var. purpureum Eschw. in Mart., Flora Brasil. 1: 170. 1833. Arthonia gregaria var. purpurea Muell.-Arg., Bull. Soc. Bot. Belg. 30: 85. 1891. Bailey (1891a); Muell.-Arg. (1891b, 1893); Shirley (1888b, 1891b). Q, V.

A. COMPLANATA Fée, Essai Crypt. Ecorc. Officin. p. 54. 1824. Muell.-Arg. (1893). V.

*A. CONSPERSULA Stirt. ex Bailey, Queensland Agr. J. 5: 39. 1899. Q.

(Arthonia)

*A. DELICATULA Muell.-Arg., Flora 65: 501. 1882. Muell.-Arg. (1891a); Shirley (1889a); Willey (1890). Q.

A. DISPERSA (Schrad.) Nyl., Lich. Scand. p. 261. 1861. Opegrapha dispersa Schrad., Ann. Bot. (Usteri) 22: 86. 1797. Muell.-Arg. (1893); Shirley (1894). V.

A. FUSCORUFA Knight ex F. Wils., Victoria Naturalist 5: 31. 1888, nomen nudum. V.

*A. GRACILENTA Muell.-Arg., Flora 65: 501. 1882. Shirley (1889a); Willey (1890). Q.

* GRACILIOR Muell.-Arg., Bull. Herb. Boiss. 3: 322. 1895. Shirley (1896). Q.

*A. GRACILLIMA Muell.-Arg., Flora 70: 424. 1887. Shirley (1889A); Willey (1890). Q.

* A. INTERSTES Muell.-Arg., Hedwigia 24: 31. 1895. V.

*A. LECIDEOLA Muell.-Arg., Bull. Herb. Boiss. 1: 60. 1893. V.

*A. LEPTOSPORA Muell.-Arg., Hedwigia 30: 53. 1891. Bailey (1891a); Shirley (1891B). Q.

A. MICROSPERMA Nyl. ex Muell.-Arg., Flora 70: 74. 1887. Muell.-Arg. (1893). V.

*A. NIGRORUFA Muell.-Arg., Bull. Herb. Boiss. 1: 59. 1893. V.

*A. NYMPHAEOIDES Knight, Trans. Linn. Soc. London, Bot. 2: 44. 1882. NSW.

A. PARDALIS F. Wils., Victoria Naturalist 6: 61. 1889, nomen nudum.

*A. PROPINQUA Nyl., Acta Soc. Scient. Fenn. 7: 484. 1863. Hue (1890-92); Muell.-Arg. (1893); Nylander (1886). NSW, V.

A. RADIATA (Pers.) Ach. f. ASTROIDEA Ach., Kongl. Svenska Vetensk.-Akad. Nya Handl. p. 131. 1808. Lichen astroites Ach., Lichenogr. Suec. Prodrom. p. 24. 1798. Arthonia vulgaris var. astroidea Schaer., Lich. Helvet. Spicil. 1: 8. 1823. Hue (1890-92); Nylander (1886); Shirley (1889A); Wilson (1888). NSW, V.

A. RUBELLA (Fée) Nyl., Mem. Soc. Sci. Nat. Cherbourg 4: 98. 1856. Graphis rubella Fée, Essai Crypt. Ecorc. Officin. p. 43. 1824. Shirley (1896). Q.

*A. SUBCONDITA Stirt. ex Bailey, Queensland Agric. J. 5: 39. 1899. Q.

A. SUBGYROSA Nyl., Flora 52: 72. 1869. Bailey (1891A); Muell.-Arg. (1891B); Shirley (1891A). Q.

*A. THOZETIANA Muell.-Arg., Flora 65: 501. 1882. Muell.-Arg. (1891A, 1893); Shirley (1889A); Willey (1890). Q.

NOTE: The following species originally described by Mueller-Arg. from the Transvaal, Africa, were reported erroneously for Australia by Willey (1890). Willey evidently thought that the locality "Transwaalia" was in Australia: A. albida, A. angulosa, A. atrorufa, A. oblongula, A. obvelata, A. pyrenuloides, A. variabilis, and A. wilmsiana.

ARTHOPYRENIA

A. ATOMARIA (Ach.) Muell.-Arg., Mém. Soc. Phys. Hist. Nat. Genève 16: 429. 1862. Lichen atomarius Ach., Lichenogr. Suec. Prodr. p. 6. 1798. Muell.-Arg. (1893a, 1895b); Shirley (1895). Q, V.

A. CINCHONAE (Ach.) Muell.-Arg., Flora 66: 287. 1883. Verrucaria cinchonae Ach., Syn. Lich. p. 90. 1814. Muell.-Arg. (1895b); Shirley (1893a). Q.

A. CONSOBRINA (Nyl.) Muell.-Arg., Flora 66: 317. 1883. Verrucaria consobrina Nyl., Ann. Sci. Nat., Bot. (4) 15: 53. 1861. Sine loc.

A. CINEREOPRUINOSA (Schaer.) Koerb., Syst. Lich. German. p. 368. 1855. Verrucaria cinereopruinosa Schaer., Lich. Helv. Spicil,., Sect. 6: 343. 1833. Muell.-Arg. (1895b). Q.

[A. DENIGRANS Muell.-Arg. was erroneously attributed to Australia by Zahlbr., Cat. Lich. 1; 303. 1921. It was collected on Victory I., between Singapore & Borneo].

*A. DIRUMPENS Muell.-Arg., Bull. Herb. Boiss. 3: 325. 1895. NSW.

* A. EXTANS Muell.-Arg., Rep. Austral. Assoc. Adv. Sci. p. 451. 1895. Shirley (1895). Q.

A. FALLACIOR Muell.-Arg., Bot. Jahrb. 6: 404. 1885. Muell.-Arg. (1895b); Shirley (1895). Q.

A. FALLAX (Nyl.) Arn., Verh. zool.-bot. Ges. Wien 23: 505. 1873. Verrucaria epidermidis var. fallax Nyl., Bot. Notiser p. 178. 1852. Nylander (1886); Hue (1890-92). NSW.

*A. GRAVASTELLA (Kremp.) Muell.-Arg., Flora 65: 518. 1882; Verrucaria gravastella Kremp., Nuovo Giorn. Bot. Ital. 7: 48. 1875. Muell.-Arg. (1883b, 1895b). Q. [Zahlbruckner, Cat. Lich. Univ. misspelled this "gravatella"]

A. LIMITANS Muell.-Arg., Flora 66: 306. 1883. Verrucaria limitans Nyl., Flora 49: 295. 1866. Muell.-Arg. (1895b); Shirley (1895). Q.

*A. OCULATA Muell.-Arg., Rep. Austral. Assoc. Adv. Sci. p. 451. 1895. Shirley (1895). Q.

*A. PICEA Shirley, Lich. Fl. Queensland 4: 174. 1890. Verrucaria picea Knight ex Shirley, l.c. Muell.-Arg. (1895b). Q.

*A. SUBOCULATA Muell.-Arg., Bull. Herb. Boiss. 3: 325. 1895. Shirley (1896). Q.

*A. STENOTHECA Muell.-Arg., op. cit. 1: 64. 1893. V.

*A. SUBPUNCTIFORMIS Muell.-Arg., Flora 71: 142. 1888. V.

*A. ZOSTRA (Knight) Shirley, Lich. Fl. Queensland 4: 174. 1890. Verrucaria zostra Knight, Trans. Linn. Soc. London, Bot. 2: 39. 1882. NSW. Hue (1890-92); Muell.-Arg. (1895b). NSW, Q.

ARTHOTHELIUM

A. AMPLIATUM (Knight) Muell.-Arg., Bull. Herb. Boiss. 1: 61. 1893. Arthonia ampliata Knight & Mitt., Trans. Linn. Soc. London 23: 106. 1860. Muell.-Arg. (1893). V.

*A. CINEREOARGENTEUM (Knight) A. Zahlbr., Cat. Lich. Univ. 2: 123. 1922. Arthonia cinereoargentea Knight ex Shirley, Proc. Roy. Soc. Queensland 6: 212. 1889. Q. Willey (1890).

*A. HUEGELII (Nyl.) A. Zahlbr., Cat. Lich. Univ. 2: 127. 1922. Arthonia huegelii Nyl., Ann. Sci. Nat., Bot. (5) 7: page? 1867. Lecidea huegelii Kremp., Verh. Zool.-Bot. Ges. Wien 18: 329. 1868. NSW. Hue (1890-92); Willey (1890).

A. INTERVENIENS (Nyl.) Muell.-Arg., Trans. Roy. Soc. Edinburgh 31: 382. 1888. Arthonia interveniens Nyl., Acta Soc. Scient. Fenn. 7: 482. 1863. Muell.-Arg. (1893). V.

A. MACROTHECUM (Fée) Mass., Ricerch. Auton. Lich. p. 55. 1852. Arthonia macrotheca Fée, Suppl. Essai Cryptog. Ecorc. Officin. p. 42. 1837. Shirley (1896). Q.

*A. MICROSPORUM Muell.-Arg., Bull. Herb. Boiss. 3: 323. 1895. Shirley (1896). Q.

*A. POLYCARPUM Muell.-Arg., Ibid., Shirley (1896). Q.

*A. PULVERULENTUM Muell.-Arg., Op. cit. 1: 60. 1893. V.

*A. PUNICEUM Muell.-Arg., Hedwigia 32: 133. 1893. Shirley (1893a). Q.

*A. VELATIUM Muell.-Arg., Bull. Herb. Boiss. 1: 60. 1893. V.

ASPICILIA

A. ALPINA (Sommerf.) Arn., Verh. zool.-bot. Ges. Wien 21: 1107. 1871. Lecanora alpina Sommerf., Suppl. Fl. Lappon. p. 91. 1826. NSW: Kosciusko State Park; trail to Northcote Pass, 7000 ft., 10 Nov. 1967, Weber & McVean L-47191 (COLO).

A. CALCAREA (L.) Mudd var. CAESIOALBA (Le Prév.) Hazsl., Magy. Birod. Zuzmoflor. p. 129. 1884. Lecanora calcarea var. caesioalba Muell.-Arg., Hedwigia 31: 196. 1892. Urceolaria caesioalba Le Prév. in Duby, Bot. Gall. 2: 671. 1830. Muell.-Arg. (1892b, 1893). V, WA.

(LECANORA) CALCAREA var. HOFFMANNII (Ach.) Sommerf., Suppl. Fl. Lappon. p. 102. 1826. Lichen hoffmannii Ach., Lichenogr. Suec. Prodrom. p. 31. 1798. Lecanora hoffmannii Muell.-Arg., Flora 72: 511. 1889. Aspicilia hoffmannii Flagey, Cat. Lich. Alger. p. 51. 1896. Muell.-Arg. (1893). V.

*(LECANORA) MACROSPERMA Muell.-Arg., Bull. Herb. Boiss. 1:40. 1893. V.

(LECANORA) VIRIDESCENS (Mass.) Muell.-Arg., loc. cit. p. 39. Pachyospora viridescens Mass., Richerch. Auton. Lich. p. 45. 1852. V.

(LECANORA) IMPRESSA (Kremp.) A. Zahlbr., Cat. Lich. Univ. 5: 319. 1928. Lecidea impressa Kremp., Flora 59: 318. 1876. Muell.-Arg. (1891a); Shirley (1891a, 1893a). Q.

(Aspicilia)

(LECANORA) IMPRESSA var. ANGULOSA (Muell.-Arg.) A. Zahlbr., Cat. Lich. Univ. 5: 320. 1928. Lecidea impressa var. angulosa Muell.-Arg., Bull. Soc. Bot. Belg. 30: 66 1891. Bailey (1881); Muell.-Arg. (1891a); Shirley (1891b). Q.

ASPIDOTHELIUM

A. FUGIENS (Muell.-Arg.) R. Sant. ex Thorold, J. Ecol. 40: 129. 1952. Lecania fugiens Muell.-Arg., Lichenes epiphylli novi, p. 3. 1890. Santesson (1952). Q.

ASTEROPORUM

A. PUNCTULIFORME (Muell.-Arg.) A. Zahlbr. in Engler & Prantl, Nat. Pflanzenfam. (2) 8: 93. 1926. Asterotrema punctuliforme Muell.-Arg., Bull. Herb. Boiss. 3: 324. 1895. Shirley (1896). Q.

*A. RIMALE Muell.-Arg., Bull. Herb. Boiss. 3: 324. 1895. NSW.

AULAXINA

A. EPIPHYLLA (A. Zahlbr.) R. Sant., Symb. Bot. Upsal. 12 (1): 305. 1952. Dictyographa epiphylla A. Zahlbr., Ann. Cryptog. Exot. 1: 123. 1928. Santesson (1952). Q.

BACIDIA

B. ALUTACEA (Kremp.) A. Zahlbr., Cat. Lich. Univ. 4: 174. 1926. Lecidea alutacea Kremp., Flora 61: 519. 1878. Shirley (1889a). Q.

B. BUCHANANII (Stirt.) Hellb., Bih. Kongl. Svenska Vetensk.-Akad. Handl. 21, afd. 3, no. 13:98. 1896. Stereocaulon buchananii Stirt., Trans. Proc. New Zealand Inst. 7:367. 1875. *Gomphillus baeomyceoides F. Wils., Victoria Naturalist 5: 30. 1888, J. Linn. Soc. London, Bot. 28: 370. 1891. V. *Patellaria wilsonii Muell.-Arg, Flora 71: 541. 1888. V. Lamb (1954); Wilson (1888, 1892). V.

*B. CAMPBELLIAE (Muell.-Arg.) A. Zahlbr., Cat. Lich. Univ. 4: 103. 1926. Patellaria campbelliae Muell.-Arg., Bull. Herb. Boiss. 4:95. 1896. Muell.-Arg., (1898). V.

*B. CONSPICUA (Knight) A. Zahlbr., Cat. Lich. Univ. 4: 107. 1926. Lecidea conspicua Knight, Trans. Linn. Soc. London, Bot. 2: 44. 1882. NSW

*B. CONVEXA (Muell.-Arg.) A. Zahlbr., Cat. Lich. Univ. 4: 188. 1882. Patellaria convexa Muell.-Arg., Flora 65: 519. 1882. NSW.

B. EFFUSA (Sm.) Trevis., Linnaea 28: 293. 1856. Lichen effusus Sm. in Sm. & Sowerby , Engl. Bot., p. 26: tab. 1863. 1808. Shirley (1889a). Q.

B. ENDOLEUCA (Nyl.) Kickx, Flore Crypt. Flandres 1: 261. 1867. Biatora luteola f. endoleuca Nyl, Bot. Notiser p. 98. 1853. Lecidea endoleuca Nyl., Acta Soc. Scient. Fenn. 7: 460. 1863. Hellbom (1896); Hue (1890-92); Nylander (1886). NSW.

*B. ENTOCOSMENSIS (Knight) A. Zahlbr., Cat. Lich. Univ. 4: 194. 1926. Lecidea entocosmensis Knight, Trans. Linn. Soc. London, Bot. 2: 44. 1882. NSW.

*B. ENTODIAPHANA (Knight) A. Zahlbr., Cat. Lich. Univ. 4: 194. 1926. Lecidea entodiaphana Knight, Trans. Linn. Soc. London, Bot. 2: 43. 1882. Shirley (1891a); Wilson (1889b). NSW.

(Bacidia)

*B. FRATRUELIS (Muell.-Arg.) A. Zahlbr., Cat. Lich. Univ. 4: 198. 1926. Patellaria fratruelis Muell.-Arg., Flora 65: 489. 1882. NSW.

*B. FURFURELLA (Muell.-Arg.) A. Zahlbr., Cat. Lich. Univ. 4: 200. 1926. Patellaria furfurella Muell.-Arg., Flora 65: 489. 1882. Lecidea furfurella Shirley, Proc. Roy. Soc. Queensland 6: 178. 1889. Muell.-Arg. (1891a). Q.

*B. INCOMPTA (Borr.) Anzi var. SPISSA (Shirley) A. Zahlbr., Cat. Lich. Univ. 4: 208. 1926. Lecidea incompta var. spissa Shirley, Proc. Roy. Soc. Queensland 6: 80. 1889. Q.

*B. LEUCOLOMA (Muell.-Arg.) A. Zahlbr., Cat. Lich. Univ. 4: 117. 1926. Patellaria leucoloma Muell.-Arg., Bull. Herb. Boiss. 1: 49. 1893. V. Filson (1965).

*B. LIVIDOFUSCA (Nyl.) A. Zahlbr., Cat. Lich. Univ. 4: 121. 1926. Lecidea lividofusca Nyl, Bull. Soc. Linn. Normand. (2) 2: 81. 1868. Lecidea lividofusca Kremp., Verh. zool.-bot. Ges. Wien 18: 328. 1868. Hue (1890-92). NSW.

*B. LIVIDONIGRANS (Muell.-Arg.) A. Zahlbr., Cat. Lich. Univ. 4: 252. 1926. Patellaria lividonigrans Muell.-Arg., Bull. Herb. Boiss. 4: 96. 1896. Muell.-Arg. (1898). V.

B. LUTEOLA (Schrad.) Mudd f. CONSPONDENS (Nyl.) A. Zahlbr., Cat. Lich. Univ. 4: 219. 1926. Lecidea luteola f. conspondens Nyl., Bull. Soc. Linn. Normand. (2) 2:86. 1868. Patellaria luteola var. conspondens Muell.-Arg., Bull. Herb. Boiss. 1: 49. 1893. Muell.-Arg. (1893); Stirton (1899b). Q, V.

B. MILLEGRANA (Tayl.) A. Zahlbr. in Wawra & Beck, Itin. Princip. Coburg. 2: 152. 1888. Lecanora millegrana Tayl., London J. Bot. 6: 159. 1847. Lecidea millegrana Nyl., Flora 41: 380. 1858. Patellaria millegrana Muell.-Arg., Flora 63: 280. 1880. Muell.-Arg. (1893); Shirley (1893a); Stirton (1899b). Q. V.

B. MILLEGRANA var. FUSCONIGRESCENS (Nyl.) A. Zahlbr., Cat. Lich. Univ. 4: 223. 1926. Lecidea millegrana var. fusconigrescens Nyl., Acta Soc. Scient. Fenn. 7: 461. 1863. Patellaria millegrana var. fusconigrescens Muell.-Arg., Flora 63: 280. 1880. Shirley (1893a). Q.

*B. MODESTULA (Muell.-Arg.) A. Zahlbr., Cat. Lich. Univ. 4: 224. 1926. Patellaria modestula Muell.-Arg., Bull. Herb. Boiss. 4: 96. 1896. Muell.-Arg. (1898). V.

*B. MULTISEPTATA (Shirley) A. Zahlbr., Cat. Lich. Univ. 4: 224. 1926. Patellaria multiseptata Shirley, Proc. Roy. Soc. Queensland 8: 134. 1892. Shirley (1892b). Q.

B. PALLIDA (Muell.-Arg.) Darb., Wiss. Ergebn. Schwed. Suedpolarexped. 1901-1903, 4 (11); 49. 1912. Patellaria pallida Muell.-Arg., Mission Scientif. Cap Horn 5: 167. 1888. Muell.-Arg. (1893). V.

B. PALLIDOCARNEA (Muell.-Arg.) A. Zahlbr., Cat. Lich. Univ. 4: 231. 1926. Patellaria pallidocarnea Muell.-Arg., Flora 64: 232. 1881. Santesson (1952). NSW.

*B. PALLIDONIGRANS (Muell.-Arg.) A. Zahlbr., Cat. Lich. Univ. 4: 134. 1926. Patellaria pallidonigrans Muell.-Arg., Bull. Herb. Boiss. 1: 49. 1893. V.

(Bacidia)

*B. QUADRILOCULARIS (Knight) A. Zahlbr., Cat. Lich. Univ. 4: 234. 1926. Lecidea quadrilocularis Knight ex Bailey, Proc. Roy. Soc. Queensland. 1: 152. 1884; Q. Patellaria contraria Muell.-Arg., Flora 70: 337. 1887. Lecidea contraria Shirley, Proc. Roy. Soc. Queensland 6: 177. 1889. Q.

B. RAFFII (Stirt.) A. Zahlbr., Cat. Lich. Univ. 4: 235. 1926. Lecidea raffii Stirt., Trans. Glasgow Soc. Field Nat. 4: 168. 1876. Bailey (1881, 1883); F. Mueller (1881); Shirley (1889a). Q.

*B. RHODOCARDIA (Muell.-Arg.) A. Zahlbr., Cat. Lich. Univ. 4: 235. 1926. Patellaria rhodocardia Muell.-Arg., Nuovo Giorn. Bot. Ital. 23: 393. 1891. Shirley (1891b). Q.

*B. RUDIS (Muell.-Arg.). A. Zahlbr., Cat. Lich. Univ. 4: 239. 1926. Patellaria rudis Muell.-Arg., Bull. Herb. Boiss. 4: 96. 1896. Muell.-Arg. (1898). Q.

*B. SEPTOSIOR (Nyl.) A. Zahlbr., Cat. Lich. Univ. 4: 240. 1926. Lecidea septosior Nyl, ex Crombie, J. Linn. Soc. London, Bot. 17: 400. 1880. Hue (1890-92). NSW.

B. SPHAEROIDES (Dicks.) A. Zahlbr., in Engler & Prantl, Nat. Pflanzenfam. I Teil, Abt. I* p. 135. 1905. Lichen sphaeroides Dicks., Fasc. Pl. Cryptog. Brit. 1: 9. 1785. Patellaria sphaeroides Muell.-Arg. Nuovo Giorn. Bot. Ital. 24: 198. 1892. Shirley (1892a). Q.

(MICROPHIALE) SPRUCEI (Bab. ex Nyl.) A. Zahlbr., Cat. Lich. Univ. 2: 701. 1924. Lecanora sprucei Bab. ex Leight., Trans. Linn. Soc. London 25: 446. 1866 (nomen nudum); Nyl., Lich. Fueg. Patagon. p. 33. 1888. Bailey (1881, 1883); F. Mueller (1881). Q. Santesson (1952, p. 475) suggests that Lecanora sprucei Bab. ex Nyl. may be Bacidia subternella (Nyl.) R. Sant. However, he does not verify its occurrence in Australia.

*B. SUBPROPOSITA (Muell.-Arg.) A. Zahlbr., Cat. Lich. Univ. 4: 244. 1926. Patellaria subproposita Muell.-Arg., Flora 65: 489. 1882. Lecidea subproposita Shirley, Proc. Roy. Soc. Queensland 6: 178. 1889. Q.

*B. SUPERBULA (Muell.-Arg.) A. Zahlbruckner, Cat. Lich. Univ. 4: 245. 1926. Patellaria superbula Muell.-Arg., Bull. Herb. Boiss. 4: 95. 1896. Muell.-Arg. (1898). V.

B. SUPERULA (Nyl.) Hellb., Bih. Kongl. Svenska Vetensk.-Akad. Handl. 21, III (13): 100. 1896. Lecidea superula Nyl., Bull. Soc. Linn. Normand. (2) 2; 87. 1868. Bailey (1881, 1883); F. Mueller (1881); Shirley (1889a). Q.

*B. TRISEPTATA (Hepp) A. Zahlbr., Cat. Lich. Univ. 4: 161. 1926. Lecidea triseptata Hepp in Zollinger, System. Verzeichn. Indisch. Archipel. ges. Pfl. p. 9. 1854. Patellaria triseptata Muell.-Arg., Nuovo Giorn. Bot. Ital. 23: 127. 1891 (overlooked by Zahlbruckner, Cat. Lich. Univ.). Bailey (1891a); Muell.-Arg. (1891b); Shirley (1891a). Q.

*B. VINICOLOR (Stirt.) A. Zahlbr., Cat. Lich. Univ. 4: 247. 1926. Lecidea vinicolor Stirt. ex Bailey, Queensland Agr. J. 5: 487. 1899. Q.

BAEOMYCES

*B. FRENCHIANUS Muell.-Arg., Flora 70: 287. 1887. Wilson (1892). V.

B. FUNGOIDES (Sw.) Ach., Method. Lich. p. 320. 1803. Lichen fungoides Sw., Nov. Gen. Sp. Pl. p. 146. 1788. Wilson (1887, 1891b, 1892). V.

*B. FUSCOCARNEUS F. Wils., Victoria Naturalist, 6: 67. 1889 ["fuscocarnea"], J. Linn. Soc. London, Bot. 28: 371. 1891. Wilson (1889a, 1891b, 1892). V.

B. HETEROMORPHUS Nyl. ex Bab. & Mitt. in Hook., Fl. Tasman. 2: 351. 1860. Wilson (1889a, 1891b, 1892). V.

B. ROSEUS Pers., Ann. Bot. (Usteri) 1: 19. 1794. Bailey (1881, 1883); Shirley (1889a); Wilson (1892). Q, V.

B. RUFUS (Huds.) Rebent., Prodrom. Fl. Neomarch. p. 315. 1804. Lichen rufus Huds., Fl. Angl. p. 443. 1762. Wilson (1892). V.

BIATORELLA

*B. AUSTRALICA Raes., Arch. Soc. Zool. Bot. Fenn. "Vanamo" 3: 184. 1949. Q.

*B. CONSANGUINEA (Stirt.) A. Zahlbr., Cat. Lich. Univ. 5: 35. 1927. Miltidea consanguinea Stirt., Trans. Proc. New Zealand Inst. 30: 385. (1897) 1898. Sine loc.

B. CONSPERSA (Fée) Vain., Etud. Lich. Brésil 2: 62. 1890. Lecidea conspersa Fée, Essai Crypt. Ecorc. Officin. p. 108. 1824. Muell.-Arg. (1892a, 1893); Shirley (1889a). Q, V.

*B. HAEMATINA (Muell.-Arg.) A. Zahlbr., Cat. Lich. Univ. 5: 41. 1927. Lecidea haematina Muell.-Arg., Hedwigia 30: 49. 1891. Q. Bailey (1891a); Shirley (1891b). Q.

BLASTENIA

"B. COCCINEA Knight." Wilson (1888, 1889b) nomen nudum.

B. ENDOCHROMOIDES (Nyl.) Muell.-Arg., Bull. Herb. Boiss. 2, appendix I, p. 69. 1894. Lecanora endochromoides Nyl., Ann. Sci. Nat., Bot. (5) 7: 326. 1867. Lecidea endochromoides Nyl., Lich. Nov. Zel. p. 89. 1888. Muell.-Arg. (1893). V.

*B. OCHROLEUCA Muell.-Arg., Bull, Herb. Boiss. 3: 641. 1895. Lecidea ochroleuca Hue, Nouv. Arch. Mus. Paris (5) 4: 16. (1912) 1914. Shirley (1896). Q.

*B. PULCHERRIMA Muell.-Arg., Flora 71: 141. 1888. Lecidea pulcherrima Hue, Nouv. Arch. Mus. Paris (5) 4: 18. (1912) 1914. V.

B. RADOMMA (Nyl. ex Knight) A. Zahlbr., Cat. Lich. Univ. 7: 40. 1930. Lecidea radomma Nyl. ex Knight, Trans. Proc. New Zealand Inst. 7: 357. 1875.

*B. SOREDIANS Muell.-Arg., Bull. Herb. Boiss. 1: 49. 1893. Lecidea soredians Hue, Nouv. Arch. Mus. Paris (5) 4: 18 (1912) 1914. V.

BOMBYLIOSPORA

*B. AUSTRALIENSIS (Muell.-Arg.) A. Zahlbr., Cat. Lich. Univ. 7: 50. 1930. Patellaria australiensis Muell.-Arg., Flora 65: 488. 1882. Lecidea australiensis Shirley, Proc. Roy. Soc. Queensland 6: 174. 1889. Muell.-Arg. (1892a). Q.

B. CHLORITIS (Tuck.) A. Zahlbr., Denkschr. K. Akad. Wiss., Math. Naturwiss. Kl. 83: 130. 1909. Lecidea chloritis Tuck. ex Nyl., Ann. Sci. Nat., Bot. (4) 11: 260. 1859. Bailey (1881). Q.

(Bombyliospora)

*B. DOLICHOSPORA (Nyl.) A. Zahlbr. var. OBSCURATA Raes., Arch. Soc. Zool. Bot. Fenn. "Vanamo" 3: 182. 1949. Q.

B. DOMINGENSIS (Pers.) A. Zahlbr. in Wawra & Beck, Itin. Princip. Coburg. 2: 153. 1888. Patellaria domingensis Pers., Ann. Wetterau. Ges. 2: 12. 1811. Lecanora domingensis Ach., Syn. Lich. p. 336. 1814. Miltidea domingensis Stirt., Trans. Proc. New Zealand Inst. 30: 386. (1897) 1898. Bailey (1881, 1883, 1891a); F. Mueller (1881); Muell.-Arg. (1882b, 1891b); Shirley (1888b, 1889a,d, 1891a); Wilson (1889b). NSW. Q.

*B. DOMINGENSIS var. CORALLOIDEA (Muell.-Arg.) A. Zahlbr., Cat. Lich. Univ. 7: 52. 1930. Patellaria domingensis var. coralloidea Muell.-Arg., Flora 65: 488. 1882. Lecidea domingensis var. coralloidea Shirley, Proc. Roy. Soc. Queensland 6: 176. 1889. Q.

(LECANORA) DOMINGENSIS var. GYROSA Bailey, Proc. Roy. Soc. Tasmania for 1879-80: 38. 1881. Nomen nudum.

B. DOMINGENSIS var. INEXPLICATA (Nyl.) Malme, Ark. Bot. 18(12): 5. 1923. Lecidea domingensis var. inexplicata Nyl., Act. Soc. Scient. Fenn. 7: 462. 1863. Patellaria domingensis var. inexplicata Muell.-Arg., Flora 69: 307. 1886. Muell.-Arg. (1891a); Shirley (1891a). Q.

*B. INFLEXA (Knight ex Bailey) A. Zahlbr., Cat. Lich. Univ. 7: 54. 1930. Lecidea inflexa Knight ex Bailey, Syn. Queensland Fl. 1. Suppl., p. 74. 1886. Shirley (1889a). Q.

B. TUBERCULOSA (Fée) Mass., Ricerch. Auton. Lich p. 116. 1852. Lecidea tuberculosa Fée, Essai Crypt. Ecorc. Officin. p. 107. 1824. Patellaria tuberculosa Spreng., Syst. Veg. 4(1): 264. 1827. Muell.-Arg. (1891a); Shirley (1891a). Q.

*B. TUBERCULOSA (Fée) Mass. var AUSTRALICA Raes., Arch. Soc. Zool. Bot. "Vanamo" 3: 182. 1949. Q.

*B. VENTRICOSA (Muell.-Arg.) A. Zahlbr., Cat. Lich. Univ. 7: 57. 1930. Patellaria ventricosa Muell.-Arg., Flora 65: 488. 1882. Lecidea ventricosa Shirley, Proc. Roy. Soc. Queensland 6: 174. 1889. Q.

BOTTARIA

B. CRUENTATA Muell.-Arg., Bot. Jahrb. 6: 395. 1885. Parmentaria astroides Fée, Suppl. Essai Cryptog. Ecorc. Officin. p. 67. 1837 (pro parte, fide Zahlbruckner). Zahlbruckner (1926, p. 84) credits this species to Australia. See Parmentaria astroidea.

*B. UMBILICATA (Knight ex Bailey) Muell.-Arg., Rep. Austral. Assoc. Adv. Sci. p. 462. 1895. Trypethelium umbilicatum Knight ex Bailey, Syn. Queensland Fl. 1. Suppl. p. 77. 1886. Parmentaria umbilicata Shirley, Lich. Fl. Queensland 4: 161. 1890. Shirley (1888b, 1895). Q.

BUELLIA

B. AETHALEA (Ach.) Th. Fr., Lichenogr. Scand. 1: 604. 1874. Lecidea atroalbella Nyl. ex. Leight., Lich. Fl. Great Brit. p. 310. 1871. Hue (1890-92); Nylander (1857). Sine loc.

B. ALBOATRA (Hoffm.) Branth & Rostr., Bot. Tidsskr. 4: 239. 1869. Lichen alboater Hoffm., Enum. Lich. p. 30. 1784. Diplotomma alboatrum Flot., 27. Jahresb. Schlesisch. Ges. vaterl. Kultur p. 130. 1849. Muell.-Arg. (1893). V.

*B. AMBLYOGONA Muell.-Arg., Bull. Herb. Boiss. 3: 641. 1895. Shirley (1896). Q.

*B. ARENARIA Muell.-Arg., Bull. Herb. Boiss 1: 52. 1893. Lamb (1955, emended descr.). V.

*B. CALLISPORA (Knight) Steiner, Bull. Herb. Boiss. (2) 7: 645. 1907. Lecidea callispora Knight, Trans. Linn. Soc. London, Bot. 2: 45. 1882. *Lecidea metaphragmia Knight, loc. cit. p. 44. Hue (1890-92); Nylander (1886). NSW.

B. CALLISPORA var. TETRAPLA (Nyl.) Steiner, Bull. Herb. Boiss. (2) 7: 645. 1907. *Lecidea tetrapla Nyl., Flora 69: 325. 1886. NSW. Buellia tetrapla Muell.-Arg., Flora 71: 139. 1888. Bailey (1891a); Hue (1890-92); Muell.-Arg. (1892b); Nylander (1886); Shirley (1892b). NSW, Q, WA.

(B. CALLISPORA) * B. TETRAPLA var. NIGROCINCTA Muell.-Arg. ex Shirley, Queensland Agr. Bull. 18: 33. 1892. Q. (overlooked by Zahlbruckner, Cat. Lich. Univ.).

B. CRETACEA Muell.-Arg., Flora 72: 512. 1889. Wilson (1890b). WA.

*B. DEMUTANS (Stirt.) A. Zahlbr., Cat. Lich. Univ. 7: 348. 1931. Lecidea demutans Stirt., Trans. Proc. Roy. Soc. Victoria 17: 71. 1881. (non Nyl. 1867). Bailey (1881, 1883); F. Mueller (1881); Shirley (1889a). Q.

*B. DESERTORUM Muell.-Arg., Hedwigia 31: 197. 1892. WA.

B. DIPLOTOMMOIDES Muell.-Arg., Flora 64: 524. 1881. Muell.-Arg. (1893). V.

*B. ENDOLEUCA Muell.-Arg., Bull. Herb. Boiss. 1: 30. 1893. V.

*B. EXTENUATA Muell.-Arg. f. ATHALLINA Muell.-Arg., Bull. Herb. Boiss. 1: 150. 1893. V.

*B. FARINULENTA Muell.-Arg., Bull. Herb. Boiss. 1: 50. 1893. V.

*B. FERTILIS Koerb., Abh. Schles. Ges. vaterl. Kultur 2: 33. 1862. Lecidea fertilis Nyl. ex Hue, Nouv. Arch. Mus. Paris (3) 3: 142. 1891. Hue (1890-92). NSW.

*B. FULIGINOSA Muell.-Arg., Bull. Herb. Boiss. 1: 50. 1893. V.

*(LECIDEA) GLOMERELLA Stirt., Queensland Agr. J. 5: 487. 1899 (non Nyl., 1861). Rinodina glomerella A. Zahlbr., Cat. Lich. Univ. 7: 520. 1931 (error). Q.

*B. GLOMERULANS (Muell.-Arg.) A. Zahlbr., Cat. Lich. Univ. 7: 464. 1931. Catolechia glomerulans Muell.-Arg., Hedwigia 31: 195. 1892. WA.

(Buellia)

*B. HALOPHILA Muell.-Arg., Bull. Herb. Boiss. 1: 52. 1893. V.

*B. HOMOPHYLIA (Knight) A. Zahlbr., Cat. Lich. Univ. 7: 366. 1931. Lecidea homophylia Knight, Trans. Linn. Soc. London, Bot. 2: 45. 1882. NSW.

*B. HOMOPHYLIA var. AMPHIBOLA (Knight) A. Zahlbr., Cat. Lich. Univ. 7: 366. 1931. Lecidea homophylia var. amphibola Knight, Trans. Linn. Soc. London, Bot. 2: 45. 1882. NSW.

*B. HOMOPHYLIA var. EMPHYTOCARPA (Knight) A. Zahlbr., Cat. Lich. Univ. 7: 366. 1931. Lecidea homophylia var. emphytocarpa Knight, Trans. Linn. Soc. London, Bot. 2: 45. 1882. NSW.

B. HYPOMELAENA Muell.-Arg., Nuovo Giorn. Bot. Ital. 21: 361. 1889. Muell.-Arg. (1893). V.

B. INNATA Muell.-Arg., Proc. Roy. Soc. Edinburgh 11: 465. 1882. Lecidea innata Shirley ex Bailey, Queensland Dept. Agr. Bull. 9: 29. 1891. Muell.-Arg. (1891a). Q.

*B. INTERSTINCTA (Nyl.) A. Zahlbr., Cat. Lich. Univ. 7: 372. 1931. Lecidea interstincta Nyl., Flora 47: 270. 1864. Hue (1890-92). Sine loc.

*B. INTURGESCENS Muell.-Arg., Hedwigia 31: 197. 1892. WA.

*B. KREMPELHUBERI A. Zahlbr., Cat. Lich. Univ. 7: 374. 1931. Lecidea exilis Kremp., Verh. zool.-bot. Ges. Wien 30: 346. 1880. Q. Buellia exilis Muell.-Arg., Flora 70: 61. 1887 (non Kremp., 1862). Bailey (1883); F. Mueller (1881); Muell.-Arg. (1887c); Shirley (1889a). Q.

B. LACTEA (Mass.) Koerb., Parerg. Lich. p. 183. 1860. Catolechia lactea Mass., Ricerch. Auton. Lich. p. 84. 1852. Lecidea lactea Shirley ex Bailey, Queensland Dept. Agr. Bull. 9: 29. 1891. Shirley (1893a). Q.

B. LAURICASSIAE (Fée) Muell.-Arg., Rev. Mycol. 9: 85. 1887. Lecidea lauricassiae Fée, Suppl. Essai Crypt. Ecorc. Officin. p. 101. 1837. Lecidea triphragmia Nyl., Act. Soc. Linn. Bordeaux 21: 387. 1856. Bailey (1881, 1883); Shirley (1891a). Q.

*B. MACROSPORA Muell.-Arg., Bull. Herb. Boiss. 1: 51. 1893. V.

*B. MACROSPOROIDES Muell.-Arg., op. cit. 3: 642. 1895. Shirley (1896). Q.

*B. MARGINULATA (Muell.-Arg.) A. Zahlbr., Cat. Lich. Univ. 7: 464. 1931. Catolechia marginulata Muell.-Arg., Hedwigia 31: 195. 1892. WA.

B. MODESTA (Kremp.) Muell.-Arg., Flora 64: 524. 1881. Lecidea modesta Kremp., Vidensk. Meddel. Naturhist. Foren. Kjoebenhavn 5: 23. (1873) 1874. B. americana A. Zahlbr., Cat. Lich. Univ. 7: 334. 1931 (nom. superfl.) Muell.-Arg. (1891a,b, 1893); Shirley (1891a). Q, V.

B. MORIOPSIS (Mass.) Th. Fr., Lich. Scand. p. 616. 1874. Catolechia moriopsis Mass., Ricerch. Auton. Lich. p. 85. 1852. Buellia atrata (Sm.) Anzi, Cat. Lich. Sondr. p. 87. 1860. Lichen atratus Sm. in Sm. & Sowerby, Engl. Bot. p. 33. 1812 (non Hedw. 1788). NSW: summit of Mt. Stillwell above Charlotte Pass, 6500 ft., 2 Feb. 1968, Weber & McVean L-49840 (COLO). First report for Southern Hemisphere.

B. PARASEMA De Not., Giorn. Bot. Ital., anno II. Tomo I, part 1: 198. 1846. B. disciformis (Fr.) Mudd, Man. Brit. Lich. p. 216. 1861. Lecidea parasema var. disciformis Th. Fr., Lichenogr. Scand. 1: 590. 1874. Crombie (1880); Fries (1846-47); Hue (1890-92); Muell.-Arg. (1893); Shirley (1889a); Stirton (1899b).

(B. PARASEMA) * B. DISCIFORMIS var. CINEREOFERRUGINEA (Knight) A. Zahlbr., Cat. Lich. Univ. 7: 352. 1931. Lecidea disciformis var. cinereoferruginea Knight ex Shirley, Proc. Roy. Soc. Queensland, 6: 182. 1889. Q.

B. PARASEMA var. RUGULOSA (Schaer.) Koerb., Syst. Lich. German. p. 228. 1855. Lecidea punctata var. rugulosa Schaer., Lich. Helvet. Spicil. 3: 154. 1828. Muell.-Arg. (1891a); Shirley (1891a). Q.

B. PARASEMA var. SAPROPHILA (Schaer.) Koerb., Syst. Lich. German. p. 228. 1855. Lecidea punctata var. saprophila Schaer., Lich. Helv. Spicil. 3: 154. 1828. Muell.-Arg. (1891a); Shirley (1891a, 1892b). Q.

B. PARASEMA var. STIGMATEA (Koerb.) Flagey, Cat. Lich. Algerie, p. 77. 1896. Buellia stigmatea Koerb., Syst. Lich. German. p. 226. 1855. Muell.-Arg. (1893). V.

B. PARASEMA var. VULGATA Th. Fr., Lichenogr. Scand. 1: 590. 1874. Muell.-Arg. (1891a, 1892a, 1893); Shirley (1891a). Q, V.

(B. PARASEMA) * B. DISCIFORMIS var. WILSONII Raes., Arch. Soc. Zool. Bot. Fenn. "Vanamo" 3: 181. 1949. Q.

*B. PEREXIGUA Muell.-Arg., Bull. Herb. Boiss. 1: 53. 1893. V.

*(LECIDEA) PLACOMORPHA Stirt., ex Bailey, Queensland Agr. J. 5: 487. 1899. Rinodina placomorpha Zahlbr., Cat. Lich. Univ. 7: 543. 1931 (error). Q.

*B. PRUINOSA Muell. Arg., Bull. Herb. Boiss. 1: 51. 1893. V.

B. PUNCTATA (Hoffm.) Mass., Ricerch. Auton. Lich. p. 81. 1852. Verrucaria punctata Hoffm., Deutschl. Fl. p. 192. 1796. Buellia myriocarpa DeNot., Giorn. Bot. Ital., anno II, parte I, tomo I: 198. 1846. Lecidea myriocarpa Roehl., Deutschl. Fl. 2(2): 35. 1813. *Trachylia exigua F. Wils., Victoria Naturalist 6: 67. 1889, J. Linn. Soc. London, Bot. 28: 370. 1891. V. Bailey (1881, 1883); F. Mueller (1881); Muell.-Arg. (1893); Shirley (1889a, 1894); Wilson (1889a, 1890a, b, 1891b, 1892). Q, V, WA.

*B. RESTITUTA (Stirt. ex Bailey) A. Zahlbr., Cat. Lich. Univ. 7: 403. 1931. Lecidea restituta Stirt. ex Bailey, Queensland Agr. J. 5: 39. 1899. Q.

*B. RIMULATA (Nyl.) A. Zahlbr., Cat. Lich. Univ. 7: 404. 1931. Lecidea rimulata Nyl., Flora 47: 270. 1864. Hue (1890-92). Sine loc.

B. RIMULOSA Muell.-Arg., Flora 71: 543. 1888. Shirley (1893a). Q.

B. SAXATILIS (Schaer.) Koerb., Syst. Lich. German. p. 228. 1855. Calicium saxatile Schaer., Naturwiss. Anzeiger, p. 35. 1821. Bailey (1881, 1883); F. Mueller (1881); Shirley (1889a); Wilson (1888). Q,V.

B. SPURIA (Schaer.) Anzi, Cat. Lich. Sondr. p. 87. 1860. Lecidea spuria Schaer., Lich. Helvet. Spicil. 3: 127. 1828. Muell.-Arg. (1892b, 1893); Willis (1953). V, WA.

B. STELLULATA (Tayl.) Mudd, Man. Brit. Lich. p. 216. 1861. Lecidea stellulata Tayl. in Mack., Fl. Hibern. 2: 118. 1836. Hue (1890-92); Krempelhuber (1880); F. Mueller (1881); Muell.-Arg. (1891a, 1892b, 1893); Nylander (1886); Shirley (1889a). NSW, Q, V, WA.

B. SUBALBULA (Nyl.) Muell.-Arg., Rev. Mycol. 2: 79. 1880. Lecidea subalbula Nyl., Bull. Soc. Linn. Normand. (2) 2: 516. 1868. Bibby & Smith (1954); Muell.-Arg. (1892b); Willis (1953). WA.

*B. SUBARENARIA Muell.-Arg., Bull. Herb. Boiss. 1: 52. 1893. V. Shirley (1893a). Q, V.

B. SUBAREOLATA Muell.-Arg., Rev. Mycol. 10: 68. 1888. Shirley (1893a). Q.

*B. SUBCONNEXA (Stirt.) A. Zahlbr., Cat. Lich. Univ. 7: 417. 1931. Lecidea subconnexa Stirt. ex Bailey, Queensland Agr. J. 5: 38. 1899. Q.

*B. SUBCORONATA (Muell.-Arg.) Malme, Ark. Bot. 21A (14): 23. 1927. Catolechia subcoronata Muell.-Arg., Hedwigia 31: 195. 1892. WA.

B. SUBDISCIFORMIS (Leight.) Vain., Etud. Lich. Brésil 1: 167. 1890. Lecidea subdisciformis Leight., Lich. Fl. Great Brit. p. 308. 1871. Bailey (1881,1883); Hue (1890-92); Muell-Arg. (1893); Nylander (1886); Shirley (1889a). NSW. Q, V.

B. SUBDISCIFORMIS var. MEIOSPERMA (Nyl.) Steiner, Verhandl. zool. bot. Ges. Wien 57: 363. 1907. Lecidea meiosperma Nyl., Ann. Sci. Nat. Bot. (4) 15: 49. 1861. Buellia meiosperma Muell.-Arg., Rev. Mycol. 9: 80. 1887. V.

*B. SUBMARITIMA Muell.-Arg., Bull. Herb. Boiss. 1: 51. 1893. V.

*B. SUBREPLETA (Stirt.) A. Zahlbr., Cat. Lich. Univ. 7: 420. 1931. Lecidea subrepleta Stirt. ex Bailey, Queensland Agric. J. 5: 39. 1899. Q.

B. SUBSTELLULANS A. Zahlbr., Cat. Lich. Univ. 7: 420. 1931. Lecidea substellulata Nyl., Flora 69: 325. 1866. NSW. Bailey (1886); Hue (1890-92); Nylander (1886); Shirley (1889a). NSW. Q.

B. TALCOPHILA (Ach.) Koerb., Syst. Lich. German. p. 230. 1855. Lecidea talcophila Ach., Lichenogr. Univ. p. 183. 1810. Karschia talcophila Koerb., Parerg. Lich. p. 460. 1865. V.

"B. TRIBACIA var. NIGROCINCTA M." Bailey (1891a, error for B. tetrapla v. nigrocincta?).

*B. VENTRICOSA Muell.-Arg., Flora 66: 79. 1883. V.

*B. WILSONIANA Muell.-Arg., Bull. Herb. Boiss. 1: 51. 1893. V.

BYSSOLOMA

B. LEUCOBLEPHARUM (Nyl.) Vain. em. R. Sant. Symb. Bot. Upsal. 12 (1): 483. 1952. Lecidea leucoblephara Nyl., Prodrom. Fl. Novo-Granat. (4) 19: 337. 1863. Byssoloma leucoblepharum Vain., Dansk Bot. Arkiv. 4: 23. 1926. Patellaria leucoblephara Muell.-Arg., Flora 64: 110. 1881. Bailey (1891a); Hue (1890-92); Muell.-Arg. (1891b); Nylander (1886); Santesson (1952); Shirley (1889a). NSW, Q.

CALICIUM

C. ABIETINUM Pers., Tentam. Dispos. Meth. Fungorum p. 59. 1797. C. curtum Turn & Borr. in Sm. & Sowerby, Engl. Bot. p. 35.: tab. 2503. 1813. Wilson (1887, 1892). V.

*C. ABIETINUM var. MINUS (F. Wils.) A. Zahlbr., Cat. Lich. Univ. 1: 590. 1922. C. curtum var. minus F. Wils., J. Linn. Soc. London, Bot. 38: 366. 1891. V.

*C. ADSPERSUM Pers. f. EUCALYPTI (F. Wils.) F. Wils. ex A. Zahlbr., Cat. Lich. Univ. 1: 592. 1922. C. roscidum Flk. var. eucalypti F. Wils. J. Linn. Soc. London, Bot. 28: 367. 1891. Wilson (1892). V.

*C. ATRONITESCENS F. Wils. ex Bailey, Queensland Dept. Agr. Bull. 7: 30. 1891. Q.

*C. AURIGERUM F. Wils., J. Linn. Soc. London, Bot. 28: 366. 1891. Wilson (1892). V.

*C. BILOCULARE F. Wils., loc. cit. p. 364. Wilson (1892). V.

*C. CAPILLARE F. Wils., ibid. Wilson (1889a, 1892). V.

*C. CHLOROSPORUM F. Wils. ex Bailey, Queensland Dept. Agr. Bull. 7: 29. 1891. Q.

*C. CONTORTUM F. Wils., Victoria Naturalist 6: 64. 1889. J. Linn. Soc. London, Bot. 28: 363. 1891. Wilson (1889a, 1891a, 1892). V. Zahlbruckner apparently overlooked the earlier publication, in English, of many of Wilson's species, in Victoria Naturalist 6: 61-69. 1889, citing only the paper of 1891, in which the same species are re-described in Latin. The 1889 paper seems to be of equal validity as a starting point.

*C. DEFORME F. Wils., Victoria Naturalist 6: 65. 1889. Wilson (1889a, 1891a, 1892) V.

*C. FULVOFUSCUM F. Wils. ex Bailey, Queensland Dept. Agr. Bull. 7: 29. 1891. Q.

C. GLEBOSUM Muell.-Arg., Flora 70: 286. 1887. Muell.-Arg. (1887b); Shirley (1892a,b). Q.

*C. GLEBOSUM var. CONCINNUM F. Wils. ex Bailey, Queensland Dept. Agr. Bull. 7: 30. 1891. Q.

*C. GLEBOSUM var. GLAUCESCENS F. Wils. ex Bailey, Queensland Dept. Agr. Bull. 7: 31. 1891. Q.

*C. GRACILLIMUM F. Wils., Victoria Naturalist 6: 65. 1889, J. Linn. Soc. London, Bot. 28: 363. 1891. Wilson (1892). V.

C. HYPERELLUM (Ach.) Ach., Method. Lich., p. 93. 1803. Lichen hyperellus Ach., Lichenogr. Suec. Prodrom. p. 85. 1798. Wilson (1892). V.

*C. HYPERELLUM var. PERBREVE F. Wils., Proc. Roy. Soc. Victoria n.s. 5: 168. 1893. V.

*C. HYPERELLUM var. VALIDIUS Knight ex Shirley, Proc. Roy. Soc. Queensland 5: 89. 1888. Q. Shirley (1889a); Wilson (1889a, 1891b, 1892). Q, V.

C. JEJUNUM F. Wils., J. Linn. Soc. London, Bot. 28: 362. 1891. C. victoriae var. jejunum F. Wils. ex Bailey, Queensland Dept. Agr. Bull. 7: 30. 1891. Shirley (1893a); Wilson (1891c). Q.

C. LENTICULARE (Hoffm.) E. Fr., Corpus Florar. Provinc. Suec. 1: 283. 1835. Trichia lenticularis Hoffm., Veget. Cryptog. 2: 16. 1790. Stirton (1899b). Q.

*C. LENTICULARE var. BULBOSUM (F. Wils.) A. Zahlbr., Cat. Lich. Univ. 1: 609. 1922. C. bulbosum F. Wils, J. Linn. Soc. London, Bot. 28: 365. 1891. V. C. quercinum var. bulbosum F. Wils., Proc. Roy. Soc. Victoria, n.s. 5: 165. 1893. Wilson (1889a, 1892). V.

*C. LENTICULARE var. CLARENSE (F. Wils.) A. Zahlbr., Cat. Lich. Univ. 1: 609. 1922. C. quercinum var. clarense F. Wils., Proc. Roy. Soc. Victoria n.s. 5: 166. 1893. V.

*C. LENTICULARE var. MICROCARPUM (F. Wils.) A. Zahlbr., Cat. Lich. Univ. 1: 610. 1922. C. quercinum var. microcarpum F. Wils., Pap. Proc. Roy Soc. Tasmania for 1892, p. 143. 1893. Wilson (1892). V.

*C. NIGRUM Schaer. var. MINUTUM Knight ex F. Wils., Victoria Naturalist 6: 65, 1889, J. Linn. Soc. London Bot. 28: 365. 1891 (non Koerb., 1863?). Wilson (1889a, 1892). V.

*C. NIVEUM (F. Wils.) F. Wils, J. Linn. Soc. London, Bot. 28: 362. 1891. Calicium pusiolum var. niveum F. Wils., Victoria Naturalist 6: 64. 1889. Wilson (1889a, 1892). V.

*C. OBCONICUM Muell.-Arg., Bull. Herb. Boiss. 4: 87. 1896. Muell.-Arg. (1898). V.

*C. OBOVATUM F. Wils., J. Linn. Soc. London, Bot. 28: 364. 1891. Wilson (1889a, 1892). V.

*C. PACHYPUS Muell.-Arg., Nuovo Giorn. Bot. Ital. 23: 385. 1891. *C. flavidum F. Wils., Victoria Naturalist 6: 64. 1889, J. Linn. Soc. London, Bot 28: 367. 1891 V. Shirley (1891b, 1894); Wilson (1889a, 1892). Q, V.

*C. PARVULUM F. Wils., Victoria Naturalist 6: 64. 1889, J. Linn. Soc. London, Bot. 28: 363. 1891. Wilson (1889a, 1892). V.

*C. PIPERATUM F. Wils., J. Linn. Soc. London, Bot. 28: 365. 1891. Wilson (1889a, 1892). V.

*C. PRAETENUE F. Wils. ex Bailey, Queensland Dept. Agr. Bull. 7: 30. 1891. Sine loc.

C. QUERCINUM var. FULVESCENS Wilson (1890b), Nomen nudum! WA.

*C. ROSEALBIDUM F. Wils., Victoria Naturalist 6: 65. 1889, J. Linn. Soc. London, Bot. 28: 364. 1891. Wilson (1889a, 1892). V.

C. SPHAEROCEPHALUM (L.) Ach., Method. Lich. p. 91. 1803. Lichen sphaerocephalus Sw., Nova Acta Acad. Upsal. 4: 245. 1784. C. trachelinum Ach., Kongl. Svenska Vetensk.-Akad. Nya Handl. p. 279. 1808. Murray (1960a); Shirley (1893a); Wilson (1887). Q, V.

*C. SPHAEROCEPHALUM var. ELATTOSPORUM (F. Wils.) A. Zahlbr., Cat. Lich. Univ. 1: 624. 1922. C. trachelinum var. elattosporum F. Wils., J. Linn. Soc. London, Bot. 28: 366. 1891. Wilson (1892). V.

*C. SPHAEROCEPHALUM var. MEIOCARPUM (F. Wils.) A. Zahlbr., Cat. Lich. Univ. 1: 624. 1922. C. trachelinum var. meiocarpum F. Wils., Proc. Roy. Soc. Victoria, n.s. 5: 167. 1893. V.

*C. SPHAEROCEPHALUM var. QUEENSLANDIAE (F. Wils.) A. Zahlbr., Cat. Lich. Univ. 1: 624. 1922. C. trachelinum var queenslandiae F. Wils. ex Bailey, Queensland Dept. Agr. Bull. 7: 30. 1891. Q.

*C. STENOSPORUM F. Wils. ex Bailey, Queensland Dept. Agr. Bull. 7: 29. 1891. Q.

C. SUBTILE Pers., Tentamen. Dispos. Method. Fungorum, p. 60. 1797. C. parietinum Ach., Kongl. Svenska Vetensk.-Akad. Nya Handl. p. 260. 1816. Stirton (1899b); Wilson (1889a). Q, V.

*C. SUBTILE Pers. var. BILOCULARE F. Wils., Victoria Naturalist 6: 65. 1889 (overlooked by Zahlbruckner, Cat. Lich. Univ.). V.

C. TRABINELLUM Ach., Method. Lich. p. 14. 1803. C. roscidulum Nyl., Herbar. Mus. Fenn. p. 78. 1859. Wilson (1892). V.

*C. TRICOLOR F. Wils., Victoria Naturalist 6:64. 1889. J. Linn. Soc. London, Bot. 28: 367. 1891. Wilson (1889a, 1892). V.

*C. VICTORIAE Knight ex F. Wils., Victoria Naturalist 6: 64. 1889, J. Linn. Soc. London, Bot. 28: 362. 1891. Wilson (1887, 1889a, 1892). V.

*C. VICTORIAE var. ALBOCARNEUM F. Wils. ex Bailey, Queensland Dept. Agr. Bull. 7: 30. 1891. Q.

*C. WILSONII Muell.-Arg., Bull. Herb. Boiss. 4: 87. 1896. Muell.-Arg. (1898). Q.

CALOPLACA

C. AURANTIACA (Lightf.) Th. Fr., Nova Acta Reg. Soc. Scient. Upsal. (3) 3: 219. 1861. Lichen aurantiacus Lightf., Fl. Scot. 2: 810. 1777. Callopisma aurantiacum Mass., Atti Imp. Regia Istit. Veneto (2) 3 (appendix III): 70. 1852. Lecanora aurantiaca Flot. ex Spreng., Jahrb. Gewaechskunde 1(1): 143. 1820. Bibby & Smith (1954); Muell.-Arg. (1892b, 1893); Nylander (1857); Shirley (1889a); Smith (1962); Stirton (1899b). Q, V, WA.

C. AURANTIACA f. DEALBATA Th. Fr., Lichenogr. Scand. 1: 178. 1871. Callopisma aurantiacum var. dealbatum Muell.-Arg., Flora 62: 166. 1879. Muell.-Arg. (1893). V.

*C. AURANTIACA var. SUBGILVA (Muell.-Arg.) A. Zahlbr., Cat. Lich. Univ. 7: 77. 1930. Callopisma aurantiacum var. subgilvum Muell.-Arg., Bull. Herb. Boiss. 1: 37. 1892. V.

*(CALLOPISMA) BALAUSTINA F. Wils., Victoria Naturalist 6: 180. 1890, nomen nudum. WA.

C. CERINA (Ehrh. ex Hoffm.) Th. Fr., Nova Acta Reg. Soc. Scient. Upsal. (3) 3: 218. 1861. Lichen cerinus Ehrh. ex Hoffm., Descr. Adumbr. Pl. Lich. 2: 62. 1789. Hue (1890-92); Muell.-Arg. (1893). V.

*C. CERINA var. MICROCARPA (Muell.-Arg.) A. Zahlbr., Cat. Lich. Univ. 7: 95. 1930. Callopisma cerinum var. microcarpum Muell.-Arg., Bull. Herb. Boiss. 3: 634. 1895. V.

C. CERINA var. ANTHRACINA (Ach.) Oliv., Flore Lich. Orne 2: 136. 1884. Lecidea anthracina Ach., Lichenogr. Univ. p. 200. 1810. Callopisma cerinum f. obscuratum Muell.-Arg., Rev. Mycol. 6: 17. 1884. Muell.-Arg. (1893). V.

(Caloplaca)

C. CINNABARINA (Ach.) A. Zahlbr. in Engler & Prantl, Nat. Pflanzenfam., I Teil, Abt. 1*, p. 228. 1907. Lecanora cinnabarina Ach., Lichenogr. Univ. p. 402. 1810. *Urceolaria tessellata Tayl., London J. Bot. 6: 158. 1847. WA. Bailey (1881, 1883); Hue (1890-92); Krempelhuber (1880); F. Mueller (1881); Muell.-Arg. (1888d, 1891a, 1893); Shirley (1889a).

C. CINNABARINA var OPACA (Muell.-Arg.) A. Zahlbr., Cat. Lich. Univ. 7: 102. 1930. Callopisma cinnabarinum var. opacum Muell.-Arg., Flora 64: 514. 1881. Muell.-Arg. (1893); Shirley (1893a). Q, V.

C. CIRROCHROA (Ach.) Th. Fr. f. LEPROSA (Lamy) A. Zahlbr., Cat. Lich. Univ. 7: 225. 1931. Lecanora cirrochroa f. leprosa Lamy, Bull. Soc. Bot. France 30: 373. 1883. Placodium cirrochroum f. leprosum Shirley, Proc. Roy. Soc. Queensland 6: 54. 1889. Q.

*C. CLAVIGERA (Stirt. ex Bailey) A. Zahlbr., Cat. Lich. Univ. 7: 108. 1931. Placodium clavigerum Stirt. ex Bailey, Queensland Agr. J. 5: 487. 1899. Q.

C. CONJUNGENS (Nyl.) A. Zahlbr., Cat. Lich. Univ. 7: 110. 1930. Lecanora conjungens Nyl., Acta Soc. Scient. Fenn. 7: 442. 1863. Callopisma conjungens Muell.-Arg., Rev. Mycol. 10: 2. 1888. Shirley (1893a). Q.

C. ELEGANS (Link) Th. Fr., Lichenogr. Scand. 1: 168. 1871. Lichen elegans Link, Ann. Naturges. 1: 37. 1791. Parmelia elegans Ach., Method. Lich. p. 193. 1803. Placodium elegans DC. in Lam. & DC., Flore Franç. ed. 3, 2: 379. 1805. Fries (1846-47); F. Mueller (1881); Wilson (1890a). V. WA.

*C. ERYTHROSTICTA (Tayl.) A. Zahlbr., Cat. Lich. Univ. 7: 116. 1930. Lecanora erythrosticta Tayl, London J. Bot. 6: 161. 1847. Callopisma erythrostictum Muell.-Arg., Flora 71: 533. 1888. Muell.-Arg. (1888d). WA.

C. FERRUGINEA (Huds.) Th. Fr., Nova Act. Reg. Soc. Scient. Upsal. (3) 3: 223. 1861. Lichen ferrugineus Huds., Fl. Angl. p. 444. 1762. Blastenia ferruginea Mass., Atti Imp. Regia Istit. Veneto (2) 3, appendix III: 102. 1852. F. Mueller (1881); Muell.-Arg. (1892b, 1893); Shirley (1889a). Q, V, WA.

C. FLAVOVIRESCENS (Wulf.) Dalla Torre & Sarnth., Die Flecht. Tirol., p. 180. 1902. Lichen flavovirescens Wulf., Schrift. Ges. Naturf. Freunde Berlin 8: 122. 1787. Callopisma flavovirescens Mass., Sched. Critic. 7: 133. 1856. Muell.-Arg. (1893). V.

*C. FULGENS Koerb., Abh. Schles. Ges. vaterl. Kultur, Abt. f. Naturwiss. 2: 31. 1862. NSW. [Zahlbruckner incorrectly treated this as a new combination for Lichen fulgens Sw. = Fulgensia. Koerber's species is a saxicolous one which he compared with C. rubelliana Koerb.]

(CALLOPISMA) FULVUM (Schwein.) Muell.-Arg., Flora 72: 144. 1889. Lecanora fulva Schwein. in Halsey, Annal. Lyc. Nat. Hist. N. Y. 1: 13. 1824. Zahlbr., Cat. Lich. Univ. 7: 275. 1931. ("Species generis dubiae"). V.

*C. GLAUCESCENS (F. Wils. ex Shirley) A. Zahlbr., Cat. Lich. Univ. 7: 239. 1931. Amphiloma glaucescens F. Wils., Proc. Roy. Soc. Queensland 6: 54. 1889. Wilson (1888). Q.

C. GRANULARIS (Muell.-Arg.) A. Zahlbr., Cat. Lich. Univ. 7: 141. 1930. Calopisma aurantiacum var. granulare Muell.-Arg., Rev. Mycol. 10: 63. 1888. Muell.-Arg. (1892b). WA.

C. GRANULOSA (Muell.-Arg.) Jatta, Sylloge Lich. Ital. p. 237. 1900. Amphiloma granulosum Muell.-Arg., Mém. Soc. Phys. Hist. Nat. Genève 16: 380. 1862. Muell.-Arg. (1893). V.

C. HOLOCARPA (Hoffm.) Wade, Lichenologist 3: 11. 1965. Verrucaria obliterata holocarpa Hoffm., Deutschl. Fl. p. 179. 1796. Callopisma aurantiacum var. holocarpum Mass., Atti Imp. Regia Istit. Veneto (2) 3, appendix III, p. 76. 1852. Caloplaca pyracea (Ach.) Th. Fr., Kongl. Svenska Vetensk.-Akad. Handl. 7 (2): 25. 1867. Parmelia cerina var. pyracea Ach., Method. Lich. p. 176. 1803. Muell.-Arg. (1893a). V.

(C. HOLOCARPA) C. PYRACEA var. PYRITHROMA (Ach.) Flagey, Rev. Mycol. 10: 130. 1888. Callopisma pyraceum var. pyrithroma Muell.-Arg., op. cit. 2: 44. 1880. Muell.-Arg. (1893). V.

*C. LATERITIA (Tayl.) A. Zahlbr., Cat. Lich. Univ. 7: 154. 1930. Lecidea lateritia Tayl., London J. Bot. 6: 149. 1847. WA.

*(CALLOPISMA) LENTICULA F. Wils., Victoria Naturalist 6: 180. 1890. WA. Nomen nudum.

*C. MICROLOBA (Muell.-Arg.) A. Zahlbr., Cat. Lich. Univ. 7: 247. 1931. Amphiloma microlobum Muell.-Arg., Bull. Herb. Boiss. 4: 93. 1896. Muell.-Arg. (1898). Q.

C. MURORUM (Hoffm.) Th. Fr., Lichenogr. Scand. 1: 170. 1871. Lichen murorum Hoffm., Enum. Lich. p. 63. 1784. Placodium murorum DC. in Lam. & DC., Flore Franc. ed. 3, 2: 378. 1805. Shirley (1889a); Willis (1953). Q, WA.

*C. MURORUM var. AREOLATA (Muell.-Arg.) A. Zahlbr., Cat. Lich. Univ. 7: 254. 1931. Amphiloma murorum var. areolatum Muell.-Arg., Hedwigia 31: 194. 1892. WA.

*C. MURORUM var. BICOLOR (Muell.-Arg.) A. Zahlbr., Cat. Lich. Univ. 7: 254. 1931. Amphiloma murorum var. bicolor Muell.-Arg., Bull. Herb. Boiss. 1: 34. 1893. V.

C. MURORUM var. MINIATA (Hoffm.) Th. Fr., Lichenogr. Scand. 1: 170. 1871. Lichen miniatus Hoffm., Enum. Lich. p. 62. 1784. Amphiloma murorum var. miniatum Koerb., Syst. Lich. German. p. 111. 1855. Muell.-Arg. (1892b). WA.

C. MURORUM var. OBLITERATA (Pers.) Jatta, Sylloge Lich. Ital. p. 238. 1900. Lichen obliteratus Pers., Ann. Bot. (Usteri) 5: 15. 1794. Amphiloma murorum f. obliteratum Koerb., Parerg. Lich. p. 48. 1859. Muell.-Arg. (1892b, 1893). V, WA.

*C. OCHROCHROA (Muell.-Arg.) A. Zahlbr., Cat. Lich. Univ. 7: 160. 1930. Callopisma ochrochroum Muell.-Arg., Bull. Herb. Boiss. 1: 36. 1892. V.

C. PHAEANTHA (Nyl.) A. Zahlbr., Cat. Lich. Univ. 7: 163. 1931. Lecanora phaeantha Nyl., Bull. Soc. Linn. Normand. (2) 2: 63. 1868. Stirton (1899b). Q.

"(PLACODIUM) PLURILOCELLARE Muell.-Arg." F. Mueller (1887a) nomen nudum.

*C. RUBENS (Muell.-Arg.) A. Zahlbr., Cat. Lich. Univ. 7: 180. 1931. Callopisma rubens Muell.-Arg., Nuovo Giorn. Bot. Ital. 23: 389. 1891. Bailey (1891); Shirley (1891b). Q.

*(CALLOPISMA VERRUCULOSA F. Wils., Victoria Naturalist 6: 180. 1890. WA, nomen nudum.

CAMPYLOTHELIUM

*C. DEFOSSUM Muell.-Arg., Nuovo Giorn. Bot. Ital. 23: 400. 1891. Bailey (1891); Muell.-Arg. (1895b); Shirley (1891a,b). Q.

*C. NITIDUM Muell.-Arg., loc. cit. p. 401. Bailey (1891); Muell.-Arg. (1895b): Shirley (1891a,b). Q.

CANDELARIA

C. FIBROSA (Fries) Muell.-Arg. var. INDICA (Vain.) Hue, Bull. Acad. Intern. Géogr. Bot. 9: 264. 1900. C. indica Vain., Philipp. J. Sci. C, 8: 99. 1913. Zahlbruckner, Cat. Lich. Univ. 6: 9. 1929. Sine loc.

C. STELLATA (Tuck.) Muell.-Arg., Flora 72: 319. 1887. Physcia candelaria var. stellata Tuck., Proc. Amer. Acad. Arts Sci. 4: 388. 1848. Muell.-Arg. (1892a). Q.

*C. XANTHOSTIGMOIDES (Muell.-Arg.) Muell.-Arg., Bull. Herb. Boiss. 1: 33. 1892. V. *Lecanora xanthostigmoides Muell.-Arg., Flora 65: 484. 1882. Muell.-Arg. (1893). NSW, V.

CATILLARIA

*C. ALBOFLAVICANS (Muell.-Arg.) A. Zahlbr., Cat. Lich. Univ. 4: 11. 1926. Patellaria alboflavicans Muell.-Arg., Hedwigia 32: 128. 1893. Shirley (1893a). Q.

*C. BANKSIAE (Muell.-Arg.) A. Zahlbr., Cat. Lich. Univ. 4: 32. 1926. Patellaria banksiae Muell.-Arg., Bull. Herb. Boiss. 1: 47. 1893. V.

*C. BRISBANENSIS Raes., Arch. Soc. Zool. Bot. Fenn. "Vanamo" 3: 182. 1949. Q.

*C. BRYOPHILA (Muell.-Arg.) A. Zahlbr., Cat. Lich. Univ. 4: 13. 1926. Patellaria bryophila Muell.-Arg., Bull. Herb. Boiss. 1: 48. 1893. V.

*C. CONFLUENS (Muell.-Arg.) A. Zahlbr., Cat. Lich. Univ. 4: 34. 1926. Patellaria confluens Muell.-Arg., Bull. Herb. Boiss. 1: 48. 1893. V.

*C. CROCEELLA (Nyl.) A. Zahlbr., Cat. Lich. Univ. 4: 35. 1926. Lecidea croceella Nyl. ex Hue, Nouv. Arch. Mus. Paris (3) 3: 110. 1891. Nylander (1857). Sine loc.

*C. DISTORTA Koerb., Abhandl. Schles. Ges. vaterl. Kultur 2: 33. 1862. (=Rivularia sp., fide Nylander, Flora 47: 267. 1864, Zahlbruckner, Cat. Lich. Univ. 4: 86. 1926). NSW.

*C. EFFUGIENS (Muell.-Arg.) A. Zahlbr., Cat. Lich. Univ. 4: 16. 1926. Patellaria effugiens Muell.-Arg., Flora 70: 336. 1887. NSW.

*C. FLAVICANS (Muell.-Arg.) A. Zahlbr., Cat. Lich. Univ. 4: 41. 1926. Patellaria flavicans Muell.-Arg., Nuovo Giorn. Bot. Ital. 23: 391. 1891. Bailey (1891); Shirley (1891b). Q.

*C. FRENCHIANA (Muell.-Arg.) A. Zahlbr., Cat. Lich. Univ. 4: 42. 1926. Patellaria frenchiana Muell.-Arg. Bull. Herb. Boiss. 4: 94. 1896. Muell.-Arg. (1898). V.

C. GLAUCONIGRANS (Tuck.) Hasse, Bryologist 12: 102. 1909. Biatora glauconigrans Tuck., Proc. Amer. Acad. Arts Sci. 12: 179. 1877. Patellaria glauconigrans Muell.-Arg., Bull. Herb. Boiss. 1: 47. 1893. V.

*C. GROSSULINA (Stirt.) A. Zahlbr., Cat. Lich. Univ. 4: 46. 1926. Lecidea grossulina Stirt., Trans. Proc. Roy. Soc. Victoria 17: 77. 1881. Sine loc.

C. INTERMIXTA (Nyl.) Arn. ex Glowacki, Verh. zool.-bot. Ges. Wien 20: 455. 1870. Lecidea intermixta Nyl., Ann. Sci. Nat. Nat., Bot. (4) 3: 161. 1855. Shirley (1893a). Q.

C. MELACLINA (Nyl.) A. Zahlbr., Cat. Lich. Univ. 4: 20. 1926. Lecidea melaclina Nyl, Lich. Nov. Zel. p. 88. 1888. Patellaria melaclina Muell.-Arg., Bull. Herb. Boiss. 1: 48. 1893. Shirley (1893a). Q, V.

*C. MELACLINOIDES (Muell.-Arg.) A. Zahlbr., Cat. Lich. Univ. 4: 20. 1926. Patellaria melaclinoides Muell.-Arg., Bull. Herb. Boiss. 4: 94. 1896. Muell.-Arg. (1898). Q.

*C. MELALOMA (Knight) A. Zahlbr., Cat. Lich. Univ. 4: 21. 1926. Lecidea melaloma Knight, Trans. Linn. Soc. London, Bot. 2: 45. 1882. NSW.

C. MELANOTROPA (Nyl.) A. Zahlbr., Cat. Lich. Univ. 4: 57. 1926. Lecidea melanotropa Nyl., J. Linn. Soc. London, Bot. 9: 255. 1865. Patellaria melanotropa Muell.-Arg., Bull. Herb. Boiss. 1: 48. 1893. V.

*C. MYCOPHILA (Muell.-Arg.) A. Zahlbr., Cat. Lich. Univ. 4: 59. 1926. Patellaria mycophila Muell.-Arg., Bull. Herb. Boiss. 4: 94. 1896. Q.

*C. PHAEOLOMA (Knight) A. Zahlbr., Cat. Lich. Univ. 4: 63. 1926. Lecidea phaeoloma Knight, Trans. Linn. Soc. London, Bot. 2: 45. 1882. NSW.

*C. POLYCARPA (Muell.-Arg.) A. Zahlbr., Cat. Lich. Univ. 4: 63. 1926. Patellaria polycarpa Muell.-Arg., Bull. Herb. Boiss. 1: 48. 1893. V.

*C. RIMOSA (Muell.-Arg.) A. Zahlbr., Cat. Lich. Univ. 4: 69. 1926. Patellaria rimosa Muell.-Arg., Bull. Herb. Boiss. 1: 48. 1893. V.

*C. SCLEROPLACA (Muell.-Arg.) A. Zahlbr., Cat. Lich. Univ. 4: 70. 1926. Patellaria sclerocarpa Muell.-Arg., Flora 65: 487. 1882. V.

C. SUBFUSCATA (Nyl.) A. Zahlbr., Cat. Lich. Univ. 4: 73. 1926. Lecidea subfuscata Nyl., Ann. Sci. Nat., Bot. (4) 11: 243. 1859. Patellaria subfuscata Muell.-Arg., Flora 65: 488. 1882. Muell.-Arg. (1893). V.

*C. SUBINTERMIXTA (Muell.-Arg.) A. Zahlbr., Cat. Lich. Univ. 4: 73. 1926. Patellaria subintermixta Muell.-Arg., Flora 65: 487. 1882. V.

*C. SUPERFLUA (Muell.-Arg.) A. Zahlbr., Cat. Lich. Univ. 4: 75. 1926. Patellaria superflua Muell.-Arg., Flora 70: 336. 1887. Muell.-Arg. (1893). NSW, V.

*C. TENUILIMBATA (Knight) A. Zahlbr., Cat. Lich. Univ. 4: 25. 1926. Lecidea tenuilimbata Knight, Trans. Linn. Soc. London, Bot., (2) 2: 46. 1882. NSW.

*C. VERRUCOSA (Muell.-Arg.) A. Zahlbr., Cat. Lich. Univ. 4: 84. 1926. Patellaria verrucosa Muell.-Arg., Bull. Herb. Boiss. 4: 94. 1896. V.

CATINARIA

C. LEUCOPLACA (DC.) A. Zahlbr., Cat. Lich. Univ. 2: 527. 1923. Patellaria leucoplaca DC. in Lam. & DC., Flore Franç. ed. 3, 2: 347. 1805. Lecidea grossa Pers. ex Nyl., Acta Soc. Linn. Bordeaux 21: 385. 1856. Hue (1890-92); Nylander (1857); Wilson (1887). V.

CETRARIA

C. ISLANDICA (L.) Ach., Method. Lich. p. 293. 1803. Lichen islandicus L., Sp. Pl. p. 1145. 1753. Bibby (1946); Weber (1969). NSW.

CHAENOTHECA

C. CHRYSOCEPHALA (Turn.) Th. Fr., Nova Acta Reg. Soc. Scient. Upsal. (3) 3: 350. 1861. Lichen chrysocephalus Turn., Trans. Linn. Soc. London, Bot. 7: 88. 1803. Calicium chrysocephalum Ach., Method. Lich. Suppl. p. 15. 1803. Wilson (1892). V.

C. CHRYSOCEPHALA f. FILARIS (Ach.) Blomb. & Forss., Enum. Pl. Scand. p. 96. 1880. Calicium chrysocephalum var. filare Ach., Kongl. Svenska Vetensk.-Akad. Nya Handl. p. 281. 1808. Wilson (1891b, 1892). V.

*C. PHAEOCEPHALA (Turn.) Th. Fr. var. PHAEDROSPORA (F. Wils.) A. Zahlbr., Cat. Lich. Univ. 1: 574. 1922. Calicium phaeocephalum var phaedrosporum F. Wils., Proc. Roy. Soc. Victoria n.s. 5: 162. 1893. V.

CHIODECTON

*C. CONGESTULUM Nyl., Bull. Soc. Linn. Normand. (2) 2: 106. 1868. Hue (1890-92); Krempelhuber (1868).

C. EFFUSUM Fée, Essai Crypt. Ecorc. Officin. p. 63. 1824. Shirley (1889a). Q.

C. ENCEPHALODES F. Wils., Victoria Naturalist 6: 61. 1889, nomen nudum. =C. grossum, fide Shirley (1894).

C. ENDOLEUCUM Muell.-Arg., Hedwigia 32: 133. 1893. Shirley (1893a). Q.

C. FARINACEUM Fée, Ann. Sci. Nat., Bot. 17: 25. 1829. Bailey (1881, 1883); Muell.-Arg. (1881); Shirley (1889a). Q.

*C. GROSSUM Muell.-Arg., Bull. Herb. Boiss. 1: 61. 1893. Shirley (1894). V.

C. HAMATUM Nyl., Ann. Sci. Nat., Bot. (4) 15: 51. 1861. Shirley (1896). Q.

*C. HETEROGENUM (Knight) A. Zahlbr., Cat. Lich. Univ. 2: 478. 1923. Stigmatidium heterogenum Knight, Trans. Linn. Soc. London, Bot. 2: 42. 1882. NSW.

*C. MACULATUM (Knight) A. Zahlbr., Cat. Lich. Univ. 2: 480. 1923. Stigmatidium maculatum Knight, Trans. Linn. Soc. London, Bot. 2: 41. 1882. NSW.

*C. NEOHOLLANDICUM Nyl., Bull. Soc. Linn. Normand (2) 2: 98. 1868. Hue (1890-92).

*C. OCHRACEOFUSCESCENS Knight ex Bailey, Syn. Queensland Fl. 1: 75. 1886. Shirley (1889a). Q.

C. SPHAERALE Ach., Syn. Lich. p. 108. 1814. *C. stromaticum Knight, Trans. Linn. Soc. London, Bot. 2: 42. 1882. Bailey (1881, 1883); Hue (1890-92); Knight (1884c); F. Mueller (1881); Muell.-Arg. (1891a, 1893); Nylander (1886); Shirley (1889a, 1893a); Wilson (1888). NSW, Q, V.

*C. STICTATHECIUM (Knight) A. Zahlbr., Cat. Lich. Univ. 2: 482. 1923. Stigmatidium stictathecium Knight, Trans. Linn. Soc. London, Bot. 2: 42. 1882. NSW.

*C. SUBDEPRESSUM Muell.-Arg., Bull. Herb. Boiss. 1: 62. 1893. V.

*C. SUBLAEVIGATUM Kremp., Verh. zool.-bot. Ges. Wien 30: 342. 1880. F. Mueller (1881); Shirley (1889a). Sine loc.

*C. TRYPETHELIOIDES (Muell.-Arg.) A. Zahlbr., Cat. Lich. Univ. 2: 482. 1923. Enterographa trypethelioides Muell.-Arg., Flora 70: 424. 1887. Chiodecton hypoleucum Knight ex Bailey, Syn. Queensland Fl. 1: 76. 1886 (non Nyl., 1855) Muell.-Arg. (1891a); Shirley (1889a). Q.

*C. VELATUM Muell.-Arg., Bull. Herb. Boiss. 1: 61. 1893. V.

*C. VIRENS Muell.-Arg., op. cit. 3: 324. 1895. Shirley (1896). Q.

CHONDROPSIS

*C. SEMIVIRIDIS (F. Muell. ex Nyl.) Nyl. ex Cromb., J. Linn. Soc. London, Bot. 17: 397. 1879. Parmeliopsis semiviridis F. Muell. ex Nyl., Syn. Lich. 2: 57. 1863. Parmelia semiviridis P. Bibby, Muelleria 1: 60. 1955. *P. hypoxantha Muell.-Arg., Flora 64: 85. 1881. SA. *P. hypoxantha var. major Muell.-Arg., op. cit. 66: 77. 1883. Q. Bibby & Smith (1954); Filson (1967); Hue (1890-92); Kurokawa (1971); F. Mueller (1881); Muell.-Arg. (1881a, 1883a, 1892b); Shirley (1889b); Tate (1882, 1887); Weber (1969); Willis (1959b). NSW, SA, Q.

CLADIA

C. AGGREGATA (Sw.) Nyl., Bull. Soc. Linn. Normand. (2) 4: 167. 1870. Lichen aggregatus Sw., Nov. Gen. Sp. Pl. p. 147. 1788. Cladonia aggregata Ach., Kongl. Svenska Vetensk.-Akad. Nya Handl. 16: 68. 1795. Clathrina aggregata Muell.-Arg., Flora 66: 80. 1883. Lichen multiflorus R. Br., Iter. Austr. No. 532. 1876. Cenomyce australis Pers., Voy. Uranie p. 213. 1826. *Cenomyce diatrypa Tayl., London J. Bot. 6: 186. 1847 (type from Macquarrie River). Cladonia cornicularia Laur. ex Flk., Clad. Comm. p. 180. 1828 (not disposed of by Vainio?). Cenomyce terebrata Laur., Linnaea 2: 43. 1827. (Muell.-Arg. proposed Clathrina because the spelling of Nylander's Cladia was too close, in his opinion, to Cladium of the Cyperaceae). Bailey (1881, 1883, 1891a); Bibby (1954); Bibby & Smith (1954). Crombie (1880); Darbishire (1912); Frey (1967); Hue (1890-92); Krempelhuber (1868, 1880); LaBillardiere (1802); F. Mueller (1881); Muell.-Arg. (1883a, 1891a, 1891b, 1892a,b); Nylander (1857); Shirley (1889a, 1892b): Smith (1962); Stirton (1899b); Tate (1881); Turner (1905); Weber 1969; Willis (1953); Wilson (1889c, 1890a); Zahlbruckner (1896, 1910). NSW, Q. SA, WA, V.

(CLADONIA) AGGREGATA f. INFLATA F. Wils., Pap. Proc. Roy. Soc. Tasmania 1892, p. 153. 1893. Wilson (1887). V.

*(CLATHRINA) AGGREGATA Muell.-Arg. var pygmaea Muell.-Arg., Bull. Herb. Boiss. 4: 88. 1896. Muell.-Arg. (1896, 1898). V.

(CLADONIA) AGGREGATA var. STRAMINEA Muell.-Arg., Flora 62: 162. 1879. Clathrina aggregata var. straminea Muell.-Arg., Nuovo Giorn. Bot. Ital. 23: 386. 1891. Shirley (1891a). Q.

*(CLADONIA) AGGREGATA f. SUBDIVERGENS Hellb., Bih. Kongl. Svenska Vetensk.-Akad. Handl. 21, Afd. III, No. 13: 89. 1896. Leighton (1866, 1867). V.

*C. CORALLAIZON Filson, Victoria Naturalist 87: 324. 1970. NSW. Weber 1969 (no. 237 sub C. retipora).

*C. FERDINANDII (Muell.-Arg.) Filson, Victoria Naturalist 87: 325. 1970. Cladonia ferdinandii Muell.-Arg., Flora 65: 293. 1882. Clathrina ferdinandii Muell.-Arg., Flora 66: 80. 1883. Cladonia retipora var. ferdinandii Vainio, Monogr. Clad. Univ. 1: 234. 1887. F. Mueller (1881). WA.

C. RETIPORA (LaBill.) Nyl., Bull. Soc. Linn. Normand. (2) 4: 167. 1870. Baeomyces reteporus LaBill., Nov. Holl. Plant. p. 110. 1804. Clathrina retipora Muell.-Arg., Flora 66: 80. 1883. Cladonia retipora Fr., Nov. Sched. Critic. 21. 1826. Lichen cribrosus R. Br., Iter Austr. no. 533, 534. 1876. Cenomyce retipora Ach., Syn. Lich. p. 248. 1814. Bailey (1881, 1883); Bibby & Smith (1954); Burnett (1925); Crombie (1880); Fries (1846-47): Hellbom (1896); Hooker (1842); Krempelhuber (1880); Laurer (1827); Leighton (1866, 1867); F. Mueller (1881); Muell.-Arg. (1891a, 1892b); Nylander (1857); Shirley (1889a); Smith (1962); Tate (1881); Turner (1905a, b); Vainio (1887); Weber 1969. NSW, Q, SA, WA, V.

*(CLADONIA) RETIPORA var. ARCUATA Stirt., Trans. Proc. New Zealand Inst. 30: 392. 1897. Stirton (1897, 1898). V. Stirton cites Scottish Naturalist p. 308. 1888 (not seen).

C. SCHIZOPORA (Nyl.) Nyl., Bull. Soc. Linn. Normand. (2) 4: 167. 1870. Cladonia schizopora Nyl., Syn. Lich. p. 217. 1860. Cladina schizopora Nyl., Flora 49: 179. 1866. Clathrina schizopora Muell.-Arg. op. cit. 66: 80. 1883. Bailey (1881); F. Mueller (1881); Weber (1969); Wilson (1887). Q, V.

C. SULLIVANII (Muell.-Arg.) W. Martin, Trans. Roy. Soc. New Zealand, Bot. 2(2): 44. 1962. Clathrina sullivanii Muell.-Arg., Flora 66: 80. 1883. Cladonia sullivanii Muell.-Arg., op. cit. 65: 294. 1882. Vainio (1887). V.

CLADONIA

C. AMAUROCRAEA (Flk.) Schaer., Lich. Helvet. Spicil. p. 34. 1823. Capitularia amaurocraea Flk., Beschr. Braunfr. Becherfl. p. 334. 1810. Hue (1890-92); Leighton (1867): Nylander (1857); Vainio (1887). WA.

C. BACILLARIS (Ach.) Nyl., Not. Saellsk. Fauna Fl. Fenn. Foerhandl. 8: 179. 1866. Baeomyces bacillaris Ach., Method. Lich. p. 329. 1803. Watts (1903); Zahlbruckner (1896). NSW.

C. CAPITELLATA (Hook. f. & Tayl.) Hook. f., Fl. Nov. Zeland. 2: 296. 1855. Cenomyce capillata (sic) Hook. F. & Tayl., London J. Bot. 3: 652. 1844. Hellbom (1896); Leighton (1867); Muell.-Arg. (1883a); Nylander (1857, as C. amaurocraea var. capitellata Bab.); Tate (1887); Vainio (1887); Zahlbruckner (1905). NSW, SA, V.

C. CARIOSA (Ach.) Spreng., Linn. Syst. Veg. ed. 16, 4: 272. 1827. Lichen cariosus Ach., Lichenogr. Suec. Prodrom. p. 198. 1798. Bibby & Smith (1954); Stirton (1899b); Wilson (1888). Q, V, WA.

*C. CARIOSA var. DIFFISSA F. Wils., Victoria Naturalist 6: 67. 1889. V. Overlooked by Vainio and Zahlbruckner.

"C. CARNEOLA var. GRACILIS?" Hampe (1853). Nomen nudum.

C. CENOTEA (Ach.) Schaer., Lich. Helvet. Spicil. p. 35. 1823. Baeomyces cenoteus Ach., Method. Lich. p. 345. 1803. Wilson (1887) V.

C. CERATOPHYLLA (Sw.) Spreng., Linn. Syst. Veg., ed. 16, 4: 271. 1827. Lichen ceratophyllus Sw., Nov. Gen. Sp. Pl. p. 147. 1788. F. Mueller (1881); Vainio (1887). NSW.

C. CHLOROPHAEA (Flk.) Spreng., Linn. Syst. Veg., ed. 16, 4: 273. 1827. Cenomyce chlorophaea Flk. in Sommerf., Suppl. Fl. Lappon. p. 130. 1826. Smith (1962); Wilson (1887). V, WA.

C. COCCIFERA (L.) Willd., Fl. Berolin. p. 361. 1787. Lichen cocciferus L., Sp. Pl. p. 1151. 1753. Laurer (1827). Sine loc.

C. COCCIFERA f. CORNUCOPIOIDES (S. Gray) Vain., Monogr. Clad. Univ. 1: 157. 1887. Scyphophora asotea beta cornucopioides S. Gray, Nat. Arr. Brit. Pl. p. 423. 1821. Cladonia cornucopioides F. Wils., Pap. Proc. Roy. Soc. Tasmania for 1892: 151. 1893. Hellbom (1896); Krempelhuber (1868, 1880); Leighton (1867); Wilson (1890a). V.

C. CONIOCRAEA (Flk.) Spreng., Linn. Syst. Veg. ed. 16. 4: 272. 1827. Cenomyce coniocraea Flk., Deutschl. Fl. 7: 11. 1821. Cladonia fimbriata var. coniocraea (Flk.) Vain., Monogr. Clad. Univ. 2: 308. 1894. C. fimbriata var. pulverulenta (Del.) Muell.-Arg., Flora 67: 619. 1884. Bailey (1891a); Muell.-Arg. (1891b); Shirley (1891a); Vainio (1894); Wilson (1887). Q, V.

C. CORALLIFERA (Kunze) Nyl., Flora 57: 70. 1874. Cenomyce corallifera Kunze in sched. descr. edita (1827). F. Mueller (1881); Vainio (1887). V.

C. CORNUTA (L.) Hoffm., Pl. Lich. 1: pl. 25. 1791. Lichen cornutus L., Sp. Pl. p. 1152. 1753. *Cladonia furcata var. notabilis Muell.-Arg. Flora 66: 18. 1883. Vainio (1894); Wilson (1887). V.

C. CRISPATA (Ach.) Flot., Merkw. Flecht. Hirschb. p. 4. 1839. Baeomyces turbinatus gamma B. crispatus Ach., Method. Lich. p. 341. 1803. Wilson (1887, 1890a). V.

C. DEFORMIS (L.) Hoffm., Deutschl. Fl., p. 120. 1796. Lichen deformis L., Sp. Pl. p. 1152. 1753. Darbishire (1912); Nylander (1857); Vainio (1887). Sine loc.

C. DEGENERANS (Flk.) Spreng., Linn. Syst. Veg., ed. 16, 4: 273. 1827. Baeomyces degenerans Flk., Berl. Magaz. 2: 283. 1807. C. trachyna Ach., Lichenogr. Univ. p. 552. 1810. Darbishire (1912); Krempelhumber (1880); Leighton (1867); F. Mueller (1881); Muell.-Arg. (1887c); Vainio (1894); Watts (1903); Wilson (1890a). NSW, V.

*C. DEGENERANS var. PLEUROCLADA Muell.-Arg., Flora 65: 295. 1882. Muell.-Arg. (1882a); Vainio (1894). V.

C. DEGENERANS var. SCABROSA Flk., Clad. Comm. p. 47. 1828. Muell.-Arg. (1892a); Vainio (1894). V.

C. DELICATA (Ehrh.) Flk., loc. cit. p. 7. Lichen delicatus Ehrh., Pl. Crypt. p. 247. 1793. Shirley (1888a, 1889a). Q.

"C. DELICATA var. SUBSQUAMOSA Nyl." Wilson (1887). V. Nomen nudum.

C. DIDYMA (Fée) Vain., Monogr. Clad. Univ. 1: 137. 1887. Scyphophorus didymus Fée, Essai Crypt. Ecorc. Officin. p. xcviii, ci. 1824. Cladonia muscigena Eschw. in Mart., Fl. Brasil. p. 262. 1833. F. Mueller (1881).

C. DILLENIANA Flk. Clad. Comm. p. 138. 1828. C. corymbescens Nyl. in Leight., Ann. Mag. Nat. Hist. (3) 18: 407. 1866. F. Mueller (1881); Muell.-Arg. (1877); Vainio (1887). NSW.

C. DILLENIANA var. EXALBIDA (Nyl.) Vain., Monogr. Clad. Univ. 1: 408. 1887. C. crispata f. exalbida Nyl, Lich. Exot. p. 249. 1859. *Cladonia fruticulosa Kremp. ex F. Mueller, Fragm. Phytogr. Austral., XI Suppl. p. 71. 1880 (nomen nudum). Muell.-Arg. (1887c). Q.

C. ENANTIA Nyl., Lich. Nov. Zel. p. 18. 1888. Watts (1903); Vainio (1887). NSW.

C. FIMBRIATA (L.) Fr., Lichenogr. Europ. Reform. p. 222. 1831. Lichen fimbriatus L., Sp. Pl. p. 1152. 1753. Cladonia ochrochlora var. ceratodes Flk., Clad. Comm. p. 77. 1828. *C. ochrochlora var. spadicea Muell.-Arg., Flora 65: 297. 1882. C. ochrochlora var. phyllostrota Flk., Clad. Comm. p. 79. 1828. Krempelhuber (1880); Maiden (1898); Muell.-Arg. (1891a); Shirley (1889a, 1891b, 1892b); Tate (1882); Wilson (1887); Vainio (1894). Q, SA, V.

C. FIMBRIATA var. ANTILOPAEA (Del. in Duby) Muell.-Arg., Flora 65: 294. 1882. Cenomyce antilopaea Del. in Duby, Bot. Gall. p. 626. 1830. Krempelhuber (1880); F. Mueller (1881); Muell.-Arg. (1891a, 1892a); Shirley (1891a, 1893a). Q.

C. FIMBRIATA var. BALFOURII (Cromb.) Vain., Monogr. Clad. Univ. 2: 239. 1894. C. balfourii Cromb., Lich. Ins. Rodrig. p. 433. 1877. C. fimbriata var. tenella Muell.-Arg., fide Vainio (1894). Wilson (1887). V.

C. FIMBRIATA var. BORBONICA (Del.) Vain., Monogr. Clad. Univ. 2: 343. 1894. Cenomyce borbonica Del. ex Nyl., Enum. Gen. Lich. p. 94. 1857. Cladonia borbonica Nyl., Exp. Lich. Nov. Caled. p. 40. 1862 pr.p. Vainio (1894). Q.

C. FIMBRIATA var. SIMPLEX (Weis) Flot. ex Vain., Monogr. Clad. Univ. 2: 256. 1894. Lichen fimbriatus var. simplex Weis, Pl. Crypt. Goetting. p. 84. 1770. Vainio (1894). Sine loc.

C. FIMBRIATA var. SUBULATA (L.) Vain., Monogr. Clad. Univ. 2: 282. 1894. Lichen subulatus L., Sp. Pl. p. 1153. 1753. Vainio (1894). V.

C. FIMBRIATA var. TUBAEFORMIS (Hoffm.) Fr., Lichenogr. Europ. Reform. p. 222. 1831. Cladonia pyxidata *Cl. tubaeformis Hoffm., Deutschl. Fl. 2: 122. 1796. Leighton (1867); Muell.-Arg. (1892a). V.

*C. FIRMA (Laur.) Kremp., Verh. zool.-bot. Ges. Wien 18: 309. 1868 (non Nyl., 1861). Cenomyce firma Laur., Linnaea 2: 44. 1827. Krempelhuber (1868); Vainio (1887). Sine loc.

C. FLOERKEANA (Fr.) Flk., Clad. Comm. p. 99. 1828. Cenomyce floerkeana Fr., Lich. Suec. Exs. no. 82. 1824. Bailey (1881, 1883); Krempelhuber (1880); Shirley (1889a); Vainio (1887); Wilson (1887). Q, V.

C. FLOERKEANA var. CARCATA (Ach.) Nyl., Lich. Scand. p. 62. 1861. Cenomyce carcata Ach., Lichenogr. Univ. p. 568. 1810. Vainio (1887). NSW.

C. FLOERKEANA var. INTERMEDIA Hepp, Lich. Europ. no. 291. 1857. Vainio (1887). NSW.

C. FOLIACEA (Huds.) Willd. var. ALCICORNIS (Lightf.) Schaer., Lich. Helvet. Spicil. p. 249. 1883. Lichen alcicornis Lightf., Fl. Scot. 2: 872. 1777. Bibby & Smith (1954); Vainio (1894). Sine loc.

C. FOLIACEA var. FIRMA (Nyl.) Vain., Monogr. Clad. Univ. 2: 400. 1894. C. alcicornis var. firma Nyl., Syn. Lich. p. 191. 1858-60. Muell.-Arg. (1892b); Vainio (1894). WA.

"C. FRAGILLIMA Kremp." Bibby & Smith (1954, as "C. fragilissima"); Wilson (1887). V. Nomen nudum.

C. FURCATA (Huds.) Schrad., Spicil. Fl. German. p. 107. 1794. Lichen furcatus Huds., Fl. Angl. p. 458. 1762. Bailey (1883); Bibby & Smith (1954); Hampe (1853); Hue (1890-92); Krempelhuber (1880); F. Mueller (1881); Shirley (1889a); Vainio (1887); Wilson (1889c). NSW, Q, V.

C. FURCATA var. ASPERATA Muell.-Arg., Flora 65: 295. 1882. Vainio (1887). NSW, V.

C. FURCATA var. PALAMAEA (Ach.) Nyl. ex Vain., Monogr. Clad. Univ. 1: 347. 1887. Baeomyces spinosus gamma B. palamaeus Ach., Method. Lich. p. 359. 1803. Cladonia furcata var. subulata Flk., Clad. Comm. p. 143. 1828. Wilson (1887). V.

C. FURCATA var. PINNATA (Flk.) Vain., Monogr. Clad. Univ. 1: 332. 1887. Cenomyce racemosa var. pinnata Flk. in Schleich., Cat. Absol. p. 47. 1821. Cladonia furcata var. foliolosa Del. in Duby, Bot. Gall. p. 623. 1830. *C. furcata var. tenuicaulis Muell.-Arg., Flora 65: 295. 1882. Q. *C. furcata var. hians Muell.-Arg., ibid. Q. Muell.-Arg. (1882a); Shirley (1893); Vainio (1887); Watts (1903); Zahlbruckner (1896). NSW, Q.

C. FURCATA var. POLYPHYLLA (Flk.) F. Wils., Pap. Proc. Roy. Soc. Tasmania for 1892: 149. 1893. C. furcata eta racemosa delta polyphylla Flk., Clad. Comm. p. 155. 1828. Wilson (1887). V.

C. FURCATA var. RACEMOSA (Hoffm.) Flk., Clad. Comm. p. 152. 1828. Cladonia racemosa Hoffm., Deutschl. Fl. p. 144. 1796. Laurer (1827); Leighton (1867); Muell.-Arg. (1887c); Wilson (1887); Vainio (1887). V.

C. FURCATA var. RECURVA Hoffm., Deutschl. Fl. 2: 115. 1796. *C. furcata var. gracillima Muell.-Arg., Flora 65: 296. 1882. NSW. *C. furcata var. cancellata Muell.-Arg., ibid. NSW. Vainio (1887); Wilson (1887). NSW.

C. FURCATA f. SURRECTA Flk., Clad. Comm. p. 154. 1828. Muell.-Arg. (1892a). V.

*C. FURCATA var. SUBSQUAMOSA Muell.-Arg., Flora 65: 296. 1882. Wilson (1887); Vainio (1887). NSW.

*C. FURCATA var. VIRGULATA Muell.-Arg., Flora 66: 18. 1883. V. Vainio (1887).

C. GRACILESCENS (Flk.) Vainio, Adj. Lich. Lapp. 1: 107. 1881. Capitularia degenerans O. C. (gracilescens) Flk., Beschr. Braunfr. Becherfl. p. 321. 1810. Vainio (1894). Sine loc.

C. GRACILIS (L.) Willd., Fl. Berol. Prodrom. p. 363. 1787. Lichen gracilis L., Sp. Pl. p. 1152. 1753. Bailey (1883); Leighton (1867); F. Mueller (1881); Shirley (1889a); Vainio (1894); Wilson (1888, 1890a). Q, V.

C. GRACILIS var. ASPERA Flk., Clad. Comm. p. 40. 1828. Bailey (1881); Vainio (1894). Q.

C. GRACILIS var. CAMPBELLIANA Vain., Monogr. Clad. Univ. 2: 113. 1894. C. campbelliana Gyelnik, Rév. Bryol. Lichenol. 6: 174. 1933. Frey (1967). NSW.

C. GRACILIS var. DILATATA (Hoffm.) Vain., Monogr. Clad. Univ. 2: 87. 1894. Vainio (1894). V.

*C. HASTATA F. Wils., Victoria Naturalist 6: 67. 1889 (not disposed of by Vainio). V.

C. LEPTOCLADA des Abb., Rév. Bryol. Lichenol. 16: 75. 1947. (includes Australian reports of C. alpestris, C. pycnoclada, C. sylvatica and C. rangiferina). Ahti (1961); Hellbom (1896); Krempelhuber (1868); Laurer (1827); Vainio (1887); Wilson (1887, 1888). V. WA.

*C. LEUCOCEPHALA Muell.-Arg., Flora 74: 110. 1891. Vainio (1894). Sine loc.

C. MACILENTA Hoffm., Deutschl. Fl. p. 126. 1796. Bailey (1881, var. seductrix Del., 1883); Darbishire (1912); Hellbom (1896); Krempelhuber (1880); F. Mueller (1881); Muell.-Arg. (1893b, var. flabellulata Muell.-Arg.); Nylander (1857, var. seductrix); Shirley (1889a,c, 1893a); Turner (1905); Wilson (1887); NSW, Q, V.

*C. MACILENTA var. FLABELLULATA Muell.-Arg., Hedwigia 32: 122. 1893. Q. Shirley (1893a); Vainio (1894).

C. MACILENTA var. SQUAMIGERA Vain., Monogr. Clad. Univ. 1: 109. 1887. C. macilenta var. seductrix Nyl., Enum. Gen. Lich. p. 97. 1857 pr.p. Bailey (1881); Nylander (1857). Q.

C. OCHROCHLORA Flk., Clad. Comm. p. 75. 1828. F. Mueller (1881); Muell.-Arg. (1892a); Vainio (1894). NSW, V.

C. PAPILLARIA (Ehrh.) Hoffm., Deutschl. Fl. 2: 117. 1796. Lichen papillaria Ehrh., Hannov. Mag. p. 218. 1783. Wilson (1887, 1891b). V.

*C. PELTASTICA (Nyl.) Muell.-Arg. f. PERTRICOSA (Kremp.) Vain., Monogr. Clad. Univ. 1: 373. 1887. C. pertricosa Kremp., Verh. zool.-bot. Ges. Wien 30: 331. 1880. Frey (1967); F. Mueller (1881); Muell.-Arg. (1887c); Weber (1969); Wilson (1887). NSW, V.

C. PITYREA (Flk.) Fries, Nov. Sched. Crit. p. 21. 1826. Capitularia pityrea Flk., Ges. Naturf. Fr. Berlin Mag. 2: 135. 1808. *Cladonia lepidula Kremp., Verh. zool.-bot. Ges. Wien 30: 331. 1880. V. *C. pergracilis Kremp., loc. cit. *C. lepidula var. foliolosa Muell.-Arg., Flora 66: 19. 1883. Q. C. pityrea var. subsquamosa (Leight.) Muell.-Arg., op. cit. 69: 253. 1886. C. pityrea var. foliolosa (Muell.-Arg.) Muell.-Arg. ibid. C. pityrea I. zwackhii b. hololepis Vain., Monogr. Clad. Univ. 2: 355. 1894. Leighton (1867); F. Mueller (1881); Muell.-Arg. (1883, 1883a, 1886, 1887c); Shirley (1889a); Vainio (1894); Watts (1903); Weber (1971). NSW, Q, V.

C. PLEUROTA (Flk.) Schaer., Enum. Lich. Eur. p. 186. 1850. Capitularia pleurota Flk., Beschr. Rothfr. Becherfl. p. 218. 1808. Wilson (1887, as C. cornucopioides var. pleurota). V.

C. PYXIDATA (L.) Hoffm., Deutschl. Fl. 2: 121. 1796. Lichen pyxidatus L., Sp. Pl. p. 1151. 1753. Cladonia conchata Nyl., Syn. Lich. p. 200. 1858-60. Bailey (1881); Hampe (1857, as C. neglecta var. lophyrea); Leighton (1867); F. Mueller (1881); Nylander (1857); Tate (1882); Wilson (1887, as var. marginalis; 1889c, 1890a,b). NSW, SA, V, WA.

C. PYXIDATA var. NEGLECTA (Flk.) Mass., Sched. Crit. p. 82. 1855. Capitularia neglecta Flk., Beschr. Braunfr. Becherfl. p. 306. 1810. Vainio (1894). Sine loc.

*C. RANGIFORMIS Hoffm. var. FILIFORMIS (Muell.-Arg.) Vain., Monogr. Clad. Univ. 1: 372. 1887. C. furcata var. filiformis Muell.-Arg., Flora 65: 296. 1882. Vainio (1887). Sine loc.

C. RANGIFORMIS var. PUNGENS (Ach.) Vain., Acta Soc. Fauna Fl. Fenn. 4: 361. 1887. Lichen pungens Ach., Lichenogr. Suec. Prodrom. p. 202. 1798. *C. narkodes Kremp. ex F. Mueller, Fragm. Phytog. Austral., Suppl. 11: 71. 1880. F. Mueller (1881); Muell.-Arg. (1887c); Vainio (1887); Wilson (1887). V.

C. RANGIFORMIS var. SOREDIOPHORA (Ny.) Vain., Monogr. Clad. Univ. 1: 368. 1887. C. furcata var. sorediophora Nyl. in Zwackh., Lich. Heidelb. p. 12. 1883. Bailey (1891a); Muell.-Arg. (1891b); Shirley (1891b). Q.

C. RIGIDA (Tayl. in Hook.) Hampe, Linnaea 2: 216. 1856. Cenomyce rigida Tayl. in Hook., Lich. Antarct. p. 652. 1844. Leighton (1867); Vainio (1887). V.

*C. SCABRIUSCULA (Del.) Sandst. var. ASPERATA (Muell.-Arg.) des Abb. ex Frey, Bot. Jahrb. 86: 243. 1967. C. furcata var. asperata Muell.-Arg., Flora 65: 295. 1882. NSW, V. Weber (1971); Wilson (1887).

C. SQUAMOSA (Scop.) Hoffm., Deutschl. Fl., 2: 125. 1796. Lichen squamosus Scop., Fl. Carniol. 2, ed. 2: 368. 1772. C. sparassa (Sm.) Hampe, Linnaea 25: 711. 1852. C. degenerans var. pleolepis Nyl., Herb. Mus. Fenn. p. 79. 1859. Bailey (1883); Hellbom (1896); Krempelhuber (1870); Laurer (1827); Leighton (1867); F. Mueller (1881); Muell.-Arg. (1887c); Shirley (1889a); Tate (1882, 1887); Vainio (1887). NSW, Q, SA, V.

C. SQUAMOSA var. GRACILENTA (Nyl.) Muell.-Arg., Nachtr. Lich. Gaz. p. 184. 1889. C. fimbriata var. gracilenta Nyl., Lich. Gaz. p. 53. 1883. Vainio (1887); Wilson (1887). V.

*C. SQUAMOSA var. PACHYPODA Muell.-Arg., Bull. Herb. Boiss. 4: 88. 1896. Muell.-Arg. (1898). SA.

*C. SQUAMOSULA Muell.-Arg., Flora 66: 19. 1883. Shirley (1889a); Vainio (1894). Q.

*C. SQUAMOSULA Muell.-Arg. f. ELEGANTULA (Muell.-Arg.) des Abb., Rév. Bryol. Lichenol. 34: 824. 1966. *C. elegantula Muell.-Arg., Flora 66: 18. 1883. NSW. Des Abbayes (1966); Muell.-Arg. (1887a); Shirley (1889b); Vainio (1894); Watts (1903). NSW, Q, V.

*C. STAUFFERI des Abb., Rév. Bryol. Lichenol. 34: 821. 1966. V. Frey (1967).

C. SUBCARIOSA Nyl., Flora 59: 560. 1876. Watts (1903). NSW.

C. SUBDIGITATA Vain., Monogr. Clad. Univ. 1: 180. 1887. Vainio (1887). V.

C. SUBSQUAMOSA (Nyl.) Vain., Acta Soc. Fauna Fl. Fenn. 4: 445. 1887. Cladonia delicata var. subsquamosa Nyl., Flora 49: 421. 1866. Darbishire (1912); Vainio (1887). Sine loc.

C. UNCIALIS (L.) Web. in Wigg., Primit. Fl. Holsat. p. 90. 1780. Lichen uncialis L., Sp. Pl. p. 1153. 1753. Krempelhuber (1868); Vainio (1887). Sine loc.

C. VERTICILLATA (Hoffm.) Schaer., Lich. Helvet. Spicil. p. 31. 1823. Cladonia pyxidata *C. verticillata Hoffm., Deutschl. Fl. 2: 122. 1796. Bibby & Smith (1954); Frey (1967); Hellbom (1896); Laurer (1827); F. Mueller (1881); Nylander (1857); Tepper (1883); Weber (1969); Wilson (1889c). NSW, SA.

C. VERTICILLATA var. CERVICORNIS (Ach.) Flk., Clad. Comm. p. 29. 1828. Lichen cervicornis Ach., Lichenogr. Suec. Prodrom. p. 184. 1798. Cenomyce verticillata var. phyllophora Sommerf., Suppl. Fl. Lapp. p. 131. 1826. Bailey (1881, 1883); Bibby (1954); Hellbom (1896); Krempelhuber (1880); Laurer (1827); F. Mueller (1881); Muell.-Arg. (1892a); Shirley (1889a); Vainio (1894); Watts (1903). NSW, Q, V.

C. VERTICILLATA var. EVOLUTA Th. Fr., Lich. Arctoi p. 149. 1860. Vainio (1894); Zahlbruckner (1896); NSW.

C. VERTICILLATA var. KREMPELHUBERI Vain., Monogr. Clad. Univ. 2: 187. 1894. Frey (1967). V.

"C. VESTITA Ach." Wilson (1888). Possibly refers to C. coccifera.

*C. XANTHOCLADA Muell.-Arg., Flora 65: 297. 1882. F. Mueller (1881, nomen nudum); Vainio (1887); Weber (1969); Wilson (1889a, 1890a). V.

CLATHROPORINA

*C. DESQUAMANS Muell.-Arg., Nuovo Giorn. Bot. Ital. 23: 402. 1891. Bailey (1891); Muell.-Arg. (1893b, 1895b); Shirley (1891b). Q.

*C. DESQUAMANS f. SOREDIIFERA Muell.-Arg., Hedwigia 32: 135. 1893. Muell.-Arg. (1895b); Shirley (1893a). Q.

C. ENDOCHRYSEA (Bab. ex Hook.) Muell.-Arg., Bull. Herb. Boiss. 2 (appendix): 93. 1894. Porina endochrysea Bab. ex Hook., Fl. New Zealand. 2: 306. 1855. P. pustulosa Kremp., Verh. zool.-Bot. Ges. Wien 26: 459. 1876. Clathroporina pustulosa Shirley, Lich. Fl. Queensland 4: 172. 1890. Muell.-Arg. (1895b); Shirley (1888b). Q.

*C. ENTEROXANTHA Shirley, Lich. Fl. Queensland 4: 171. 1890. Muell.-Arg. (1895b). Q.

*C. FARINOSA (Knight ex Bailey) A. Zahlbr., Cat. Lich. Univ. 1: 417. 1922. Porina farinosa Knight ex Bailey, Syn. Queensland Fl., 1. Suppl. p. 73. 1886. *Clathroporina tomentella Muell.-Arg., Flora 70: 428. 1887. Q. Muell.-Arg. (1895b); Shirley (1890). Q.

*C. FLAVESCENS Muell.-Arg., Nuovo Giorn. Bot. Ital. 23: 403. 1891. Bailey (1891); Muell.-Arg. (1895b); Shirley (1891b). Q.

*C. MEIOSPORA Shirley, Lich. Fl. Queensland 4: 172. 1890. Muell.-Arg. (1895b). Q.

*C. OLIVACEA Muell.-Arg., Flora 65: 518. 1882. Muell.-Arg. (1891a, 1892a, 1895b); Shirley (1890, 1893a). Q.

*C. ROBUSTA Muell.-Arg., op. cit. 70: 428. 1887. Muell.-Arg. (1895b); Shirley (1890). Q.

COCCOCARPIA

C. CILIOLATA Mont., Ann. Sci. Nat., Bot. (3) 10: 129. 1848. Pannaria ciliolata Hue, Bull. Soc. Bot. France 48: lx. (1901) 1902. Stirton (1899b). Q.

C. CRONIA (Tuck.) Vain. var. AURANTIACA (Hook. f. & Tayl.) Vain., Ann. Acad. Sci. Fenn. Ser. A, 15(6): 24. 1921. Solorina aurantiaca Hook. f. & Tayl., London J. Bot. 3: 635. 1844. Coccocarpia aurantiaca Mont. & v. d. Bosch in Jungh., Pl. Jungh. fasc. IV: 465. 1855. Shirley (1889a, 1892b, 1893a). Q.

C. CRONIA var. ISIDIOPHYLLA (Muell.-Arg.) Vain., Ann. Acad. Sci. Fenn. Ser. A, 6(7): 103. 1915. Coccocarpia pellita var. isidiophylla Muell.-Arg., Flora 65: 321. 1882. Bailey (1891a); Muell.-Arg. (1891b); Shirley (1891b). Q.

*C. LEUCORRHIZA Hampe, Linnaea 28: 217. 1856. F. Mueller (1881). V.

C. PELLITA (Ach.) Muell.-Arg., Flora 65: 320. 1882. Parmelia pellita Ach., Lichenogr. Univ. p. 468. 1810. Shirley (1889a). Q.

C. PELLITA var. COCOËS (Fée) A. Zahlbr., Cat. Lich. Univ. 3: 286. 1925. Circinaria cocoës Fée, Essai Crypt. Ecorc. Officin. p. 127. 1824. *Coccocarpia pellita var. semiincisa Muell.-Arg., Flora 65: 321. 1882 (SA). Shirley (1893a, 1894). Q, V.

*C. PELLITA var. MESOMORPHA Muell.-Arg., Bull. Herb. Boiss. 4: 92. 1898. Muell.-Arg. (1898). Q.

C. PELLITA var. PARMELIOIDES (Hook. in Kunth) Muell.-Arg., Rev. Mycol. 9: 139. 1887. Lecidea parmelioides Hook. in Kunth, Syn. Pl. Aequin. Orb. Nov. 1: 15. 1822. Coccocarpia molybdaea Pers. in Gaudich., Voy. Uranie, Bot. p. 206. 1826. Pannaria molybdaea Tuck., Gen. Lich. p. 52. 1872. Stirton (1899b). Q.

C. PELLITA var. SMARAGDINA (Pers.) Muell.-Arg., Flora 65: 320. 1882. C. smaragdina Pers. in Gaudich., Voy. Uranie, Bot. p. 206. 1826. Bailey (1881, 1883); Krempelhuber (1880); F. Mueller (1881); Muell.-Arg. (1887c, 1891b); Shirley (1889a, 1893a). NSW, Q.

COELOCAULON

C. ACULEATUM (Schreb.) Gyeln., Acta Faun. Fl. Fenn. (2) 1 (5-6): 7. 1933. Lichen aculeatus Schreb., Spicil. Flor. Lipsiens., p. 125. 1771. Cornicularia aculeata Ach., Method. Lich. p. 302. 1803. Fries (1846-47); F. Mueller (1881). SA, WA. The type species of Cornicularia was C. normoerica (Gunn.) Du Rietz. This differs in many fundamentals from the rest of those species usually placed in Cornicularia.

COENOGONIUM

*C. BOTRYOSUM Knight ex Bailey, Syn. Queensland Fl. 1 Suppl. p. 74. 1886. Shirley (1889a). Q.

C. CONFERVOIDES Nyl., Flora 41: 380. 1858. Bailey (1881); Shirley (1893a). Q.

C. IMPLEXUM Nyl., Ann. Sci. Nat., Bot. (4) 16: 92. 1862. Bibby (1954); Hellbom (1896); Hue (1890-92); Muell.-Arg. (1891a); Nylander (1857); Shirley (1891b). Q.

C. INTERPLEXUM Nyl., ibid. Muell.-Arg. (1893); Shirley (1889a,d). Q, V.

C. INTERPOSITUM Nyl., loc. cit. p. 91. Bailey (1881, 1883); F. Mueller (1881); Muell.-Arg. (1891a); Shirley (1889a). Q.

C. LINKII Ehrenb. ex Nees ab Esenb., Horae Phys. Berolin. p. 120. 1820. Bailey (1881, 1883, 1891a); Muell.-Arg. (1891b); Shirley (1888b, 1889a); Wilson (1887). Q, V.

C. MONILIFORME Tuck., Proc. Amer. Acad. Arts Sci. 5: 416. 1862. Biatorinopsis torulosa Muell.-Arg., Rév. Mycol. 10: 114. 1888. Muell.-Arg. (1889); Shirley (1891a). Q.

*C. ORNATUM Muell.-Arg., Bull. Herb. Boiss. 4: 96. 1896. Muell.-Arg. (1898). Q.

*C. RIGIDULUM Muell.-Arg., Flora 65: 490. 1882. Shirley (1889a). Q.

COLLEMA

C. COCCOPHORUM Tuck., Proc. Amer. Acad. Arts Sci. 5: 385. 1862. *C. atrum F. Wils., Victoria Naturalist 6: 62. 1889. (V). Degelius (1954); Shirley (1894); Wilson (1889a, 1890b, 1892). V, WA.

*C. CONGESTUM F. Wils., Victoria Naturalist 6: 61. 1889, J. Linn. Soc. London, Bot. 28: 355. 1891. Synechoblastus congestus F. Wils., Proc. Roy. Soc. Victoria (2) 5: 153. 1892. Wilson (1889a). V.

C. FURVUM (Ach.) DC. in Lam. & DC., Flor. Franç. ed. 3, 2: 385. 1805. Lichen furvus Ach., Lichenogr. Suec. Prodrom. p. 132. 1798. Shirley (1889a); Wilson (1887, 1892). Q, V.

*C. GALACTINUM (Muell.-Arg.) A. Zahlbr., Cat. Lich. Univ. 3: 38. 1924. Synechoblastus galactinus Muell.-Arg., Flora 61: 488. 1878. NSW.

*C. GWYTHERI Stirt. ex Bailey, Queensland Agr. J. 5: 484. 1899. Q.

C LAEVE Hook. f. & Tayl., London J. Bot. 3: 656. 1844. Synechoblastus laevis Muell.-Arg., Flora 70: 283. 1887. *Leptogium olivaceum F. Wils., Victoria Naturalist 6: 62. 1889. J. Linn. Soc. London, Bot. 28: 356. 1891. V. *Collema senecionis F. Wils., Victoria Naturalist 6: 62. 1889, J. Linn. Soc. London, Bot. 28: 356. 1891. V. Krempelhuber (1880); F. Mueller (1881); Muell.-Arg. (1887c); Shirley (1894); Wilson (1889a, 1892). NSW, V.

*C. LAEVE f. FIMBRIATUM F. Wils., Proc. Roy. Soc. Victoria, n.s. 5: 152. 1892. V.

*C. LAEVE f. GRANULATUM (F. Wils.) F. Wils., Proc. Roy. Soc. Victoria, n.s. 5: 152. 1892. Leptogium olivaceum var. granulatum F. Wils., Jour. Linn. Soc. London, Bot. 28: 357. 1891. V.

*C. LAEVE f. ISIDIOSUM (F. Wils.) F. Wils., Proc. Roy. Soc. Victoria, n.s. 5: 152. 1892. Leptogium olivaceum var. isidiosum F. Wils., Victoria Naturalist 6: 62. 1889. Jour. Linn. Soc. London, Bot. 28: 357. 1891. Wilson (1889a). V.

*C. LAEVE var. LIMBATUM (F. Wils.) A. Zahlbr., Cat. Lich. Univ. 3: 40. 1924. Leptogium olivaceum var. limbatum F. Wils., J. Linn. Soc. London, Bot. 28: 357. 1891. V.

C. LEUCOCARPUM Hook. f. & Tayl., London. J. Bot. 3: 657. 1844. Collema glaucophthalmum (Bab.) Nyl., Flora 41: 377. 1858. Synechoblastus leucocarpus Muell.-Arg., Flora 65: 293. 1882. S. glaucophthalmus Shirley ex Bailey, Queensland Dept. Agr. Bull. 9: 20. 1891. Crombie (1880); Krempelhuber (1880); F. Mueller (1881); Muell.-Arg. (1887b,c); Shirley (1889a); Turner (1905); Weber (1969); Willis (1953); Wilson (1887, 1889c, 1891b, 1892). Q, V, WA.

*C. LEUCOCARPUM var. MINUS F. Wils., Victoria Naturalist 6: 62. 1889. J. Linn. Soc. London., Bot. 28: 356. 1891. Synechoblastus leucocarpus var. minor F. Wils., Proc. Roy. Soc. Victoria n.s. 5: 155. 1892. Wilson (1889a,c). V.

*C. MICROCARPOIDES A. Zahlbr., Cat. Lich. Univ. 3: 41. 1924. Synechoblastus microcarpus Muell.-Arg., Flora 65: 292. 1882. Collema microcarpum Shirley, Lich. Fl. Queensland 4: 183. 1890 (non DC.). Bailey (1883); F. Mueller (1881); Shirley (1889a). Q.

C. NIGRESCENS (Huds.) DC. in Lam. & DC., Flor. Franç. ed. 3, 2: 384. 1805. Lichen nigrescens Huds., Fl. Anglica, p. 450. 1762. Hellbom (1896); Shirley (1889a); Willis (1953); Wilson (1887, 1889b,c, 1891b, 1892). NSW, Q, V, WA.

*C. PLUMBEUM F. Wils., J. Linn. Soc. London, Bot. 28: 354. 1891. Wilson (1889a, 1892). V.

C. ROBILLARDII (Muell.-Arg.) Stizenb., Ber. ueber die Thaetigk. St. Gall. Naturw. Ges. 1888-89, p. 119. 1890. Synechoblastus robillardii Muell.-Arg., Flora 60: 471. 1877. *Collema quadriloculare F. Wils., Victoria Naturalist 6: 61. 1889, J. Linn. Soc. London, Bot. 28: 356. 1891. V. Shirley (1894); Wilson (1889a, 1892). V.

CONIOCYBE

*C. CITRIOCEPHALA F. Wils., Victoria Naturalist 6: 66. 1889; J. Linn. Soc. London, Bot. 28: 368. 1891. Wilson (1892). V.

*C. GRACILENTA (Ach.) Ach. f. leucocephala F. Wils., Proc. Roy. Soc. Victoria, n.s. 5: 170. 1893. V.

*C. OCHROCEPHALA F. Wils., J. Linn. Soc. London, Bot. 28: 368. 1891. Wilson (1889a, 1892). V.

*C. RHODOCEPHALA F. Wils., Victoria Naturalist 6: 66. 1889. J. Linn. Soc. London, Bot. 28: 368. 1891. Wilson (1889c, 1892). V.

CROCYNIA

*(AMPHILOMA) GLAUCESCENS Wils., Victoria Naturalist 5: 29. 1888. Q.

C. GOSSYPINA (Sw.) Mass., Atti. Imp. Regia Istit. Veneto (3) 5: 252. 1860. Lichen gossypinus Sw., Nova Gen. Sp. Pl. p. 146. 1788. Shirley (1189a,e). Q.

CYPHELIUM

*C. BUELLIACEUM (Muell.-Arg.) A. Zahlbr. (as "buellianum"), Cat. Lich. Univ. 1: 663. 1922. Acolium buelliaceum Muell.-Arg., Hedwigia 32: 121. 1893. Shirley (1893a). Q.

*C. EMERGENS (F. Wils.) A. Zahlbr., Cat. Lich. Univ. 1: 664. 1922. Trachylia emergens F. Wils., Victoria Naturalist 6: 66. 1889, J. Linn. Soc. London, Bot. 28: 369. 1891. V. Wilson (1889a, 1892). SA, V.

(Cyphelium)

*C. LECANORINUM (F. Wils.) A. Zahlbr., Cat. Lich. Univ. 1: 667. 1922. Trachylia lecanorina F. Wils., Victoria Naturalist 6: 66. 1889, J. Linn. Soc. London, Bot. 28: 369. 1891. V. *Acolium subocellatum Muell.-Arg., Hedwigia 32: 121. 1893. Q. Shirley (1894); Wilson (1889a, 1892). Q, V.

*C. TRICINCTUM (F. Wils.) A. Zahlbr., Cat. Lich. Univ. 1: 676. 1922. Trachylia tricincta F. Wils. ex Bailey, Queensland Dept. Agr. Bull. 7: 31. 1891. Q.

*C. VICTORIANUM (F. Wils.) A. Zahlbr., Cat. Lich. Univ. 1: 676. 1922. Trachylia victoriana F. Wils., Victoria Naturalist 6: 67. 1889, J. Linn. Soc. London, Bot. 28: 370. 1891. V. Wilson (1889a, 1892). V.

*C. VIRIDILOCULARE (F. Wils.) A. Zahlbr., Cat. Lich. Univ. 1: 676. 1922. Trachylia viridilocularis F. Wils., Victoria Naturalist 6: 66. 1889, J. Linn. Soc. London, Bot. 28: 369. 1891. V. Acolium parasema Muell.-Arg., Hedwigia 32: 121. 1893. Q. Shirley (1894); Wilson (1889a, 1892). Q, V.

DERMATINA

D. PYRENOCARPA (Nyl.) A. Zahlbr., Cat. Lich. Univ. 1: 550. 1922. Mycoporum pyrenocarpum Nyl., Flora 41: 381. 1858. Mycoporum pycnocarpum Nyl., Ann. Sci. Nat., Bot. (4) 20: 242. 1863. Muell.-Arg. (1893). V.

DERMATOCARPON

D. CINEREUM (Pers.) Th. Fr., Nova Acta Reg. Soc. Scient. Upsal. (3) 3: 356. 1861. Endocarpon cinereum Pers., Ann. Bot. (Usteri) 1: 28. 1794. NSW. Northcote Pass, Kosciusko State Park, 7000 ft., 10 Nov. 1967, Weber & McVean L-47254, 42066 (COLO).

D. HEPATICUM (Ach.) Th. Fr., Nova Acta Reg. Soc. Scient. Upsal. (3) 3: 355. 1861. Endocarpon hepaticum Ach., Kongl. Vetensk.-Akad. Nya Handl. p. 156. 1809. Endopyrenium hepaticum Koerb., Parerg. Lich. p. 302. 1863. Muell.-Arg. (1892b, 1893a); Wilson (1887). V, WA.

D. MINIATUM (L.) Mann, Lich. Bohem. obs. Dispos. p. 66. 1825. Lichen miniatus L., Sp. Pl., p. 1149. 1753. Muell.-Arg. (1892a, 1895b); Shirley (1895). Q.

DICTYONEMA

D. IRPICINUM Mont., Ann. Sci. Nat., Bot. (3) 10: 119. 1848. Dichonema irpicinum Nyl., op. cit. (4) 11: 240. 1859. Bailey (1891a); Bibby (1954); Muell.-Arg. (1891a,b); Shirley (1889a,e). Q.

D. SERICEUM (Sw.) Berk., London J. Bot. 2: 639. 1843. Thelephora sericea Sw., Fl. Ind. Occid. 3: 1928. 1809. Dichonema sericeum Mont. in Bel., Voy. Ind.-Or. 2: 155. 1846. Bailey (1891a); Muell.-Arg. (1891b); Shirley (1891a, 1892b). Q.

DIMERELLA

D. EPIPHYLLA (Muell.-Arg.) Malme, Ark. Bot. 26A (13): 9. 1934. Biatorinopsis epiphylla Muell.-Arg., Flora 64: 103. 1881. Santesson (1952). NSW.

D. LUTEA (Dicks.) Trev., Rendic. Istit. Lombardo 13: 65. 1880. Lichen luteus Dicks., Fasc. Pl. Crypt. Brit. 1: 11. 1785. Biatorinopsis lutea Muell.-Arg., Flora 64: 102. 1881. F. Mueller (1881); Muell.-Arg. (1891a, 1892a, 1893); Shirley (1889a). Q, V.

(MICROPHIALE) PLANELLA (Nyl.) A. Zahlbr., Cat. Lich. Univ. 2: 700. 1924. Lecidea planella Nyl., Bull. Soc. Linn. Normand. (2) 2: 84. 1868. Knight (1889c); Shirley (1889a). Q.

*D. ZONATA (Muell.-Arg.) R. Sant., Symb. Bot. Upsal. 12(1): 399. 1952. Biatorinopsis zonata Muell.-Arg., Lich. Epiphylli Novi, p. 16. 1890. Microphiale zonata A. Zahlbr., Cat. Lich. Univ. 2: 702. 1924. Muell.-Arg. (1890a); Shirley (1891b). Q.

DIPLOGRAMMA

*D. AUSTRALIENSE Muell-Arg., Nuovo Giorn. Bot. Ital. 23: 399. 1891. Bailey (1891); Shirley (1891b). Q.

DIPLOSCHISTES

D. ACTINOSTOMUS (Pers.) A. Zahlbr., Hedwigia 31: 34. 1892. Urceolaria actinostoma Pers., ex Ach., Lichenogr. Univ. p. 288. 1810. Muell.-Arg. (1891a, 1893); Shirley (1891a, 1893a, 1894). Q, V.

D. ALBISSIMUS (Ach.) Dalla Torre & Sarnth., Die Flechten Tirol, p. 299. 1902. Urceolaria scruposa var. albissima Ach., Method. Lich. p. 147. 1803. Diploschistes scruposus var. cretaceus Muell.-Arg., Hedwigia 31: 157. 1892. Muell.-Arg. (1893). V.

D. BRYOPHILUS (Ehrh.) A. Zahlbr., Hedwigia 31: 34. 1892. Lichen bryophilus Ehrh., Pl. Crypt. Exsicc. no. 236. 1774. Urceolaria scruposa var. bryophila Ach., Method Lich. p. 148. 1803. Diploschistes scruposus var. bryophilus Muell.-Arg., Bull. Herb. Boiss. 1: 41. 1892. Muell.-Arg. (1893); Wilson (1887). V.

D. SCRUPOSUS (Schreb.) Norm., Nyt Mag. Naturvidensk. 7: 232. 1853. Lichen scruposus Schreb., Spicil. Fl. Lips. p. 133. 1771. Bibby & Smith (1954); F. Mueller (1881); Muell.-Arg. (1893); Shirley (1894); Tate (1882); Willis (1959b). SA, V, WA.

D. SCRUPOSUS var. ARENARIUS (Schaer.) Muell.-Arg., Hedwigia 33: 196. 1892. Urceolaria scruposa var. arenaria Schaer., Lich. Helvet. Spicil. 2: 75. 1826. Muell.-Arg. (1892b, 1893). V.

*D. SCRUPOSUS var. AUSTRALICA [sic] Raes., Arch. Soc. Zool. Bot. Fenn. "Vanamo" 3: 184. 1949. V.

*D. STICTICUS (Koerb.) Muell.-Arg., Bull. Herb. Boiss. 2, appendix 1: 52. 1894. Urceolaria stictica Koerb., Abhandl. Schles. Ges. vaterl. Kultur 2: 32. 1862. NSW.

D. SUBOCELLATUS (Nyl. ex Cromb.) A. Zahlbr., Cat. Lich. Univ. 2: 674. 1924. Urceolaria subocellata Nyl. ex Cromb., J. Linn. Soc. London, Bot. 17: 399. 1879. SA: soil crusts, Koonamore Vegetation Reserve ca. 45 mi NW of Yunta, 18 Nov. 1967, Weber & McVean L-47166 (COLO); WA: 54 mi W of Coolgardie on Perth-Kalgoorlie road, 21 Oct. 1967, G. C. Bratt 67/51 (COLO).

DIRINASTRUM

*D. AUSTRALIENSE Muell.-Arg., Bull. Herb. Boiss. 1: 54. 1893. V.

ECHINOPLACA

E. EPIPHYLLA Fée, Essai Crypt. Ecorc. Officin. p. xciii. 1824. Santesson (1952). Q.

E. LEUCOTRICHOIDES (Vain.) R. Sant. ex Thorold, J. Ecol. 40: 129. 1952. Calenia leucothrichoides Vain., Ann. Acad. Sci. Fenn. ser. A, 15: 166. 1921. Santesson (1952). Q.

(Echinoplaca)

E. PELLICULA (Muell.-Arg.) R. Sant., Symb. Bot. Upsal. 12(1): 367. 1952. Arthonia pellicula Muell.-Arg., Bot. Jahrb. 4: 56. 1883. Santesson (1952). Q.

ENDOCARPON

*C. HELMSIANUM Muell.-Arg., Hedwigia 31: 197. 1892. Muell.-Arg. (1893a). WA.

E. PUSILLUM Hedw., Descr. Adumbr. Muscor. Frondos. 2: 56. 1789. SA: Nullarbor Plain, Jan. 1966, McVean 6624, L-42008. NSW: 5 mi. E Cooma, 2 Oct. 1967, Weber & McVean L-49880; Big Hole near Krawaree, 19 Mar. 1968, Weber & McVean L-49709 (COLO).

*E. VICTORIAE Muell.-Arg., Bull. Herb. Boiss. 1: 62. 1893. V.

ENTERODICTYON

E. VELATUM (Knight) A. Zahlbr., Cat. Lich. Univ. 2: 470. 1923. Stigmatidium velatum Knight, Trans. Linn. Soc. London, Bot. 2: 41. 1882. Graphis develatula Nyl., Flora 69: 327. 1886. Enterodictyon knightii Muell.-Arg., Bull. Herb. Boiss. 3: 323. 1895. NSW.

ENTEROGRAPHA

*E. DIVERGENS (Muell.-Arg.) Redinger, Fedde Repert. 43: 62. 1939. Chiodecton divergens Muell.-Arg., Bull. Herb. Boiss. 1: 62. 1893. V.

EPHEBE

*E. EPHEBOIDES (F. Wils.) A. Henssen, Symb. Bot. Upsal. 18: 54. 1963. Stigonema epheboides F. Wils., J. Linn. Soc. London, Bot. 28: 354. 1891. Crombie (1879); Shirley (1894); Wilson (1892, as E. pubescens). NSW, V.

E. FRUTICOSA A. Henssen, Symb. Bot. Upsal. 18: 53. 1963. NSW: Mt. Kosciusko, 6500 ft., 5 Mar. 1968, 4 Jan. 1968, Weber & McVean L-47489, 47911 (COLO). Henssen mentions this species as occurring intermixed in the type collection of E. tasmanica.

*E. TASMANICA Cromb., J. Linn. Soc. London, Bot. 17: 391. 1880. Crombie (1880) lists this and E. pubescens, which Henssen (1963) does not recognize in the Australian flora. Henssen (1963); Hue (1890-92). The type locality (Grose River) is in New South Wales, in the Blue Mountains north of Katoomba, not in Tasmania as interpreted by Henssen and evidently Crombie, who was responsible for the specific epithet.

ERIODERMA

*E. KNIGHTII Shirley, Bailey's Bot. Bull. 8: 96. 1893. Platysma eriophyllum Knight ex Shirley, Proc. Roy. Soc. Queensland 5: 110. 1888. Cetraria eriophylla A. Zahlbr., Cat. Lich. Univ. 6: 287. 1929. Shirley (1889a). Q.

EVERNIA

E. MESOMORPHA Nyl. f. ESOREDIOSA Muell.-Arg., Flora 74: 110. 1891. Zahlbruckner (1896). NSW.

FULGENSIA

F. BRACTEATA (Hoffm.) Jatta, Sylloge Lich. Ital. p. 236. 1900. Psora bracteata Hoffm., Deutschl. Fl. p. 169. 1796. Placodium fulgens var. bracteatum Duby, Bot. Gallic. 2: 662. 1830. *Lecanora drummondii Tayl., London J. Bot. 6: 160. 1847 (WA). Placodium drummondii Nyl., Mem. Soc. Imp. Sci. Nat. Cherbourg 5: 111. 1857. Hue (1890-92); Muell.-Arg. 1888d, 1892b, 1893); Nylander (1857); Willis (1959b). V, WA.

GLYPHIS

G. CICATRICOSA Ach., Syn. Lich. p. 107. 1814. Glyphis favulosa Ach., ibid. Shirley (1889a). Q.

G. CICATRICOSA var. CONFLUENS (Zenk.) A. Zahlbr., Akad. Wiss. Wien, Math.-naturwiss. Kl., Denkschr. 83: 112. 1909. Glyphis confluens Zenk. in Goebel & Kunze, Pharmazeut. Waarenkunde 1: 163. 1827-29. Bailey (1881, 1883); F. Mueller (1881); Shirley (1889a). Q.

*G. CICATRICOSA var. DEPAUPERATA (Muell.-Arg.) A. Zahlbr., Cat. Lich. Univ. 2: 456. 1923. Glyphis favulosa var. depauperata Muell.-Arg., Hedwigia 30: 54. 1891. Bailey (1891a); Shirley (1891b). Q.

G. CICATRICOSA var. INTERMEDIA (Muell.-Arg.) A. Zahlbr., Cat. Lich. Univ. 2: 456. 1923. Bailey (1891a); Muell.-Arg. (1891a, b); Shirley (1891a). Q.

G. CRIBROSA Fée, Nova Actorum Caes. Acad. Leop.-Carol. Nat. Cur. 18, Suppl. I, p. 36. 1841. Bailey (1891a); Muell.-Arg. (1891b); Shirley (1891b). Q.

*G. VERRUCULOSA A. Zahlbr., Cat. Lich. Univ. 2: 457. 1923. Glyphis verrucosa Knight ex Shirley, Proc. Roy. Soc. Queensland 6: 214. 1889 (non Mont. & v. d. Bosch, 1855). Shirley (1889a). Q.

GRAPHINA

*G. BRACHYSPORA Muell.-Arg., Flora 66: 80. 1883. Q.

G. FISSOFURCATA (Leight.) Muell.-Arg., op. cit. 65: 385. 1882. Graphis fissofurcata Leight., Trans. Linn. Soc. London, Bot. 27: 177. 1869. Bailey (1881, 1883); F. Mueller (1881); Shirley (1889a). Q.

G. GLAUCODERMA (Nyl.) Muell.-Arg., Bull. Herb. Boiss. 3: 47. 1895. Graphis glaucoderma Nyl. ex Tuck., Synops. N. A. Lich. 2: 124. 1888. Shirley (1889a); Wilson (1887). Q, V.

*G. GYRIDIA (Stirt.) A. Zahlbr., Cat. Lich. Univ. 2: 408. 1923. Graphis gyridia Stirt., Trans. Proc. Roy. Soc. Victoria 17: 77. 1881. F. Mueller (1881); Stirton (1881b). V.

*G. HARTMANNIANA Muell.-Arg., Flora 65: 502. 1882. Shirley (1889a). Q.

G. LEPROCARPA (Nyl.) A. Zahlbr., Cat. Lich. Univ. 2: 412. 1923. Graphis leprocarpa Nyl., Acta Soc. Scient. Fenn. 7: 472. 1863. Hue (1890-92); Nylander (1886). NSW.

*G. PALMICOLA Muell.-Arg., Flora 70: 402. 1887. Shirley (1889a). Q.

*G. PERTENELLA (Stirt.) Shirley, Proc. Roy. Soc. Queensland 6: 208. 1889. Graphis pertenella Stirt., Trans. Proc. Roy. Soc. Victoria 17: 72. 1881. Bailey (1881, 1883); F. Mueller (1881). Q.

G. PLATYCARPA (Eschw.) A. Zahlbr., Akad. Wiss. Wien, Sitzungber., Math.-Naturwiss. Kl. 111: 385. 1902. Graphis platycarpa Eschw. in Mart., Fl. Brasil. 1: 74. 1833. Graphis sophistica Nyl., Ann. Sci. Nat., Bot. (4) 19: 359. 1863. Crombie (1880); Hellbom (1896); Muell.-Arg. (1891a, 1893); Shirley (1889a). NSW, Q, V.

(Graphina)

*G. PLATYCARPA var. RECTA (Muell.-Arg.) A. Zahlbr., Cat. Lich. Univ. 2: 419. 1923. Graphina sophistica var. recta Muell.-Arg., Bull. Herb. Boiss. 3: 321. 1895. Shirley (1896). Q.

*G. POLYCLADES (Kremp.) Muell.-Arg., Flora 65: 502. 1882. Graphis polyclades Kremp., Verh. zool.-bot. Ges. Wien 30: 341. 1880. F. Mueller (1881); Muell.-Arg. (1887c); Shirley (1889a). NSW, Q.

*G. REPLETA (Stirt.) Shirley, Proc. Roy. Soc. Queensland 6: 208. 1889. Graphis repleta Stirt., Trans. Proc. Roy. Soc. Victoria 17: 73. 1881. Bailey (1881, 1883); F. Mueller (1881); Shirley (1888b). Q.

*G. SAXICOLA Muell.-Arg., Flora 70: 401. 1887. Shirley (1889a,b). Q.

G. SIMULANS (Leight.) Muell.-Arg., Hedwigia 30: 52. 1891. Graphis simulans Leight., Trans. Linn. Soc. London 27: 177. 1869. Bailey (1891a); Shirley (1891a). Q.

*G. SUBAGGREGANS Muell.-Arg., Bull. Herb. Boiss. 1: 58. 1893. V.

*G. SUBTARTAREA Muell.-Arg., Flora 70: 402. 1887. Shirley (1889a,b). Q.

*G. SUBVELATA (Stirt.) A. Zahlbr., Cat. Lich. Univ. 2: 428. 1923. Graphis subvelata Stirt. ex Bailey, Queensland Dept. Agr. Bull. 5: 488. 1899. Q.

*G. TENUIRIMA Shirley, Queensland Dept. Agr. Bull. 18: 34. 1902. Q. Overlooked by Zahlbruckner, Cat. Lich. Univ.

GRAPHIS

G. AFZELII Ach., Syn. Lich. p. 85. 1814. Bailey (1881, 1883); F. Mueller (1881); Shirley (1889a). Q.

*G. ALBISSIMA Muell.-Arg., Bull. Herb. Boiss. 3: 319. 1895. Shirley (1896). Q.

*G. ALBONITENS Muell.-Arg., Hedwigia 30: 53. 1891. Muell.-Arg. (1891b); Shirley (1891b). Q.

"G. AMPLICATA Ach." Bailey (1891a). Nomen nudum.

G. ANFRACTUOSA (Eschw.) Eschw. in Mart., Fl. Brasil. 1: 86. 1833. Scaphis anfractuosa Eschw., Syst. Lich. p. 25. 1824. Muell.-Arg. (1893). V.

*G. ARGOPHOLIS Knight ex Muell.-Arg., Flora 70: 401. 1877. Shirley (1889a,b). Q.

*G. AULACOTHECIA Knight, Trans. Linn. Soc. London, Bot. 2: 41. 1882. NSW.

*G. BAILEYANA Muell.-Arg., Hedwigia 32: 132. 1893. Shirley (1893a). Q.

*G. CENTRIFUGA Raes., Arch. Soc. Zool. Bot. Fenn. "Vanamo" 3: 187. 1949. NSW.

*G. CIRCUMFUSA Stirt., Trans. Proc. Roy. Soc. Victoria 17: 73. 1881. Bailey (1881, 1883); F. Mueller (1881); Shirley (1889a). Q.

G. COMPARILIS Nyl. var. INSIDIOSA (Knight & Mitt.) Hook. f., Handb. New Zealand Flora p. 586. 1867. Fissurina insidiosa Knight & Mitt., Trans. Linn. Soc. London 23: 102. 1860. Graphis insidiosa Knight & Mitt. ex Hook f., Handb. New Zealand Flora p. 586. 1867. Shirley (1893a). Q.

*G. CRASSILABRA Muell.-Arg., Flora 65: 502. 1882. Shirley (1889a). Q.

G. DESCISSA Muell.-Arg., Bull. Herb. Boiss. 3: 318. 1895. Shirley (1896). Q.

G. DUMASTII (Fée) Spreng., Syst. Veg. 4(1): 254. 1827. Fissurina dumastii Fée, Essai Crypt. Ecorc. Officin. p. xc. 1824. V.

G. DUPLICATA Ach., Syn. Lich. p. 81. 1814. Muell.-Arg. (1891b); Shirley (1889a, 1893a). Q.

G. DUPLICATA var. PERUVIANA (Fée) A. Zahlbr., Cat. Lich. Univ. 2: 303. 1923. Opegrapha peruviana Fée, Essai Crypt. Ecorc. Officin. p. 27. 1824. Graphis duplicata var. sublaevis Muell.-Arg., Mem. Soc. Phys. Hist. Nat. Geneve 29 (8): 35. 1887. Shirley (1893a). Q.

G. ELEGANS (Sm.) Ach., Syn. Lich. p. 85. 1814. Opegrapha elegans Sm. in Sm. & Sowerby, Engl. Bot., 26: tab. 1812. 1807. Wilson (1888). V.

*G. EMERSA Muell.-Arg., Hedwigia 32: 132. 1893. Shirley (1893a). Q.

*G. EPIMELAENA Muell.-Arg., Bull. Herb. Boiss. 3: 319. 1895. Shirley (1896). Q.

*G. GLAUCA Muell.-Arg., op. cit. 1: 58. 1893. V.

*G. IMMERSELLA Muell.-Arg., op. cit. 3: 319. 1895. Shirley (1896). Q.

*G. INNATA Knight ex Shirley, Proc. Roy. Soc. Queensland 6: 201. 1889. Q.

G. INTRICATA Fée, Essai Crypt. Ecorc. Officin. p. 42. 1824. Graphis assimilis Nyl., Bull. Soc. Linn. Normand. (2) 2: 109. 1868. Bailey (1891a); Crombie (1880); Muell.-Arg. (1891a,b, 1893); Shirley (1889a, 1894). Q, V.

*G. LAEVIGATA Muell.-Arg., Nuovo Giorn. Bot. Ital. 23: 398. 1891. Bailey (1891); Shirley (1891b). Q.

G. LINEOLA Ach., Lichenogr. Univ. p. 264. 1810. Shirley (1893a). Q.

G. NIGRIRIMIS Muell.-Arg., Bull. Herb. Boiss. 3: 320. 1895. Fissurina comparilis f. nigririmis Nyl., Lich. Nov. Zel. p. 125. 1888. Muell.-Arg. (1895c). V.

G. PERSULCATA Stirt., Proc. Philos. Soc. Glasgow 11: 315. 1879. Bailey (1881, 1883); F. Mueller (1881); Shirley (1889a). Q.

*G. PROPINQUA Muell.-Arg., Flora 65: 502. 1882. Shirley (1889a). Q.

G. RIMULOSA (Mont.) Trevis., Spighe e Paglie, p. 11. 1853. Opegrapha rimulosa Mont., Ann. Sci. Nat., Bot. (2) 18: 271. 1842. Shirley (1889a). Q.

*G. RIMULOSA var. BRACHYCARPA Muell.-Arg. ex Shirley, Queensland Dept. Agr. Bull. 13: 27. 1896. Q.

G. RIMULOSA var. PULVERULENTA (Nyl.) Muell.-Arg., Bull. Soc. Bot. Belg. 30: 79. 1891. G. striatula var. pulverulenta Nyl., Flora 41: 381. 1858. Bailey (1891a); Shirley (1891a). Q.

*G. ROBUSTIOR Muell.-Arg., Nuovo Giorn. Bot. Ital. 24: 398. 1891. Bailey (1896); Shirley (1891b). Q.

*G. SAYERI Muell.-Arg., Flora 70: 401. 1887. Bailey (1891a); Muell.-Arg. (1891a,b); Shirley (1889a). Q.

G. SCRIPTA (Wigg.) Ach., Kongl. Svenska Vetensk.-Akad. Nya Handl. p. 145. 1809. Verrucaria scripta Wigg., Primit. Flor. Holsat. p. 86. 1780. Bibby & Smith (1954); Harmand (1912); Hue (1890-92); Nylander (1886); Shirley (1889a). NSW, Q, WA.

G. SCRIPTA var. PULVERULENTA (Pers.) Ach., Syn. Lich. p. 82. 1814. Opegrapha pulverulenta Pers., Ann. Bot. (Usteri) 1: 29. 1794. Muell.-Arg. (1893). V.

G. SCRIPTA var. SERPENTINA (Ach.) Meyer, Nebenstud. p. 194. 1825. Lichen serpentinus Ach., Lichenogr. Suec. Prodrom. p. 25. 1798. Hue (1890-92); Muell.-Arg. (1893); Nylander (1886). NSW, V.

G. SCRIPTA var. TYPOGRAPHICA (Willd.) A. Zahlbr., Cat. Lich. Univ. 2: 350. 1923. Verrucaria typographica Willd., Fl. Berol. Prodrom. p. 370. 1787. Graphis scripta f. recta Hepp, Flecht. Europ. no. 46. 1853. Muell.-Arg. (1891a); Shirley (1891b). Q.

*G. SEMIAPERTA Muell.-Arg., Nuovo Giorn. Bot. Ital 23: 397. 1891. Bailey (1891); Shirley (1891b). Q.

G. STRIATULA (Ach.) Spreng., Syst. Veg. 4 (1): 250. 1827. Opegrapha striatula Ach., Syn. Lich. p. 74. 1814. Shirley (1889a). Q.

*G. STRIATULA var. SUBLAEVIS Nyl., Flora 41: 381. 1858; Muell.-Arg., Bull. Herb. Boiss. 1: 57. 1893. V.

*G. SUBTENELLA Muell.-Arg. Flora 70: 400. 1887. Shirley (1889a). Q.

G. TENELLA Ach., Syn. Lich. p. 81. 1814. Shirley (1889a, 1894). Q,V.

*G. VERMIFERA Muell.-Arg., Flora 70: 401. 1887. Shirley (1889a). Q.

*G. VINOSA Muell.-Arg., Bull. Herb. Boiss. 3: 318. 1895. Shirley (1896). Q.

*G. WILSONIANA Muell.-Arg., op. cit. 1: 57. 1893. V.

*G. XANTHOSPORA Muell.-Arg., op. cit. 3: 320. 1895. Shirley (1896). Q. Zahlbruckner, Cat. Lich. Univ. 2: 361. 1923, erroneously lists this as G. "xanthocarpa."

GYALECTA

*G. LEPTOSPORA (Muell.-Arg.) A. Zahlbr., Cat. Lich. Univ. 2: 712. 1924. Secoliga leptospora Muell.-Arg., Bull. Herb. Boiss. 1: 43. 1893. V.

GYALECTIDIUM

G. FILICINUM Muell.-Arg., Flora 64: 101. 1881. G. phyllocharis Muell.-Arg., op. cit. 73: 188. 1890. Muell.-Arg. (1891a); Santesson (1952); Shirley (1893a). Q.

GYMNOGRAPHA

*G. MEDUSULINA Muell.-Arg., Flora 70: 62. 1887. Muell.-Arg. (1887a,c). Q.

HAEMATOMMA

H. BABINGTONII Mass., Bull. Soc. Imp. Naturalistes Moscou 36: 260. 1863. Knight (1884c); Shirley (1889a,c,d); Stirton (1899b); Wilson (1889c). Q, V.

*H. PHAEOPLACUM (Stirt. ex Bailey) A. Zahlbr., Cat. Lich. Univ. 5: 768. 1928. Lecanora phaeoplaca Stirt. ex Bailey, Queensland Agr. J. 5: 487. 1899. Q.

H. PUNICEUM (Ach.) Mass., Atti Imp. Regia Ist. Veneto (3) 5: 253. 1860. Lecanora punicea Ach., Lichenogr. Univ. p. 395. 1810. Parmelia punicea Ach., Method. Lich. p. 167. 1803. Lecania punicea Muell.-Arg., Rév. Mycol. 9: 134. 1887. Bailey (1881, 1883); Bibby & Smith (1954); Crombie (1880); Darbishire (1912); Fries (1846-47); Hue (1890-92); Knight (1882); F. Mueller (1881); Muell.-Arg. (1891a); Nylander (1857, 1886); Shirley (1889a). NSW, Q, WA.

*H. PUNICEUM var. COLLATUM (Stirt.) A. Zahlbr., Cat. Lich. Univ. 5: 770. 1928. Lecanora punicea var. collata Stirt., Trans. Glasgow Soc. Field Naturalists 5: 216. 1877. Bailey (1899). Q.

*H. PUNICEUM var. INFUSCUM (Stirt.) A. Zahlbr., Cat. Lich. Univ. 5: 770. 1928. Lecanora punicea var. infusca Stirt. ex Bailey, Queensland Agr. J. 5: 38. 1899. Q.

HELMINTHOCARPON

*H. BAILEYANUM Muell.-Arg. ex Bailey, Queensland Agr. Dept., Bull. 13: 29. 1896. Shirley (1896). Q.

*H. LOJKANUM Muell.-Arg., Flora 70: 423. 1887. Shirley (1889a,b). Q.

HEPPIA

*H. ACAROSPOROIDES Muell.-Arg., Hedwigia 31: 194. 1892. Hue (1907). WA.

*H. AUSTRALIENSIS Muell.-Arg., loc. cit., p. 193. Hue (1907). WA.

*H. BRISBANENSIS F. Wils. ex Bailey, Queensland Dept. Agr. Bull. 7: 32. 1891. Q.

HERPOTHALLON

H. SANGUINEUM (Sw.) Tobler, Flora 131: 447. 1937. Byssus sanguinea Sw., Nov. Gen. Sp. Pl. p. 148. 1788. Chiodecton rubrocinctum Nyl., Bot. Zeitung 20: 278. 1862. Shirley (1889a). Q.

HETERODEA

*H. MUELLERI (Hampe) Nyl., Bull. Soc. Linn. Normand (2)2: 47. 1868. Sticta muelleri Hampe, Linnaea 25: 711. 1852. ("South Coast"). Platysma muelleri Nyl., Syn. Lich. 1: 306. 1860. *Trichocladia baileyi Stirt., Trans. Proc. Roy. Soc. Victoria 18: 1. 1882. Q. Bailey (1883, 1891a); Krempelhuber (1868, 1880); F. Mueller (1881): Muell.-Arg. (1891a,b, 1892b); Shirley (1889a); Stirton (1881a, 1889b); Tate (1881, 1883); Turner (1905); Wilson (1887); Zahlbruckner (1896, 1903). NSW, Q, SA, V, WA.

HYPOGYMNIA

H. BILLARDIERI (Mont.) Filson, Victoria Naturalist 87: 325. 1970. Cetraria billardieri Mont., Syllog. Gen. Sp. Cryptog. p. 322. 1856. Parmelia billardieri (Mont.) A. Zahlbr., Cat. Lich. Univ. 6: 26. 1929.

P. placorhodioides Nyl., Syn. Lich. 1: 401. 1860. Bitter (1901); Crombie (1880); Hue (1890-92); Krempelhuber (1870); Muell.-Arg. (1883a); Nylander (1857); Stirton (1899b); Wilson (1887, 1889c). NSW, Q, V.

*(PARMELIA) CAMPBELLII Knight, Proc. Roy. Soc. Queensland 1: 114. 1884. NSW, V.

H. ENCAUSTA (Sm.) Wats., Trans. Bot. Soc. Edinburgh 32: 504. 1939. Parmelia encausta (Sm.) Ach., Method. Lich. p. 202. 1803. Lichen encaustus Sm., Trans. Linn. Soc. London 1: 83. 1791. Stirton (1899b); Wilson (1890a). Q.

H. ENTEROMORPHA (Ach.) Nyl., Acta Soc. Scient. Fenn. 26(10): 7. 1900. Parmelia enteromorpha Ach., Method. Lich. p. 252. 1803. Bitter (1901); Darbishire (1912); Hampe (1853); Hue (1890-92). V.

H. LUGUBRIS (Pers.) Krog, Norsk Polarinstitutt Skr. 144: 99. 1968. Parmelia lugubris Pers. in Gaudich., Voy. Uranie, Bot. p. 196. 1826. Weber (1967). ACT, NSW.

(PARMELIA) LUGUBRIS f. TENUIS Bitter, Hedwigia 40: 244. 1901. Sine loc.

*H. MUNDATA (Nyl.) Rassad., Bot. Mat. Gerb. Bot. Inst. Komarova Akad. Nauk USSR. 11: 11. 1956. Parmelia mundata Nyl., Syn. Lich. 1: 401. 1860. *P. physodes var. mundata Muell.-Arg., Flora 66: 176. 1883. Q. Bailey (1883); Bitter (1901); Krempelhuber (1880); F. Mueller (1881); Muell.-Arg. (1883a, 1887c); Nylander (1857); Shirley (1889a); Wilson (1887); Zahlbruckner (1896). NSW, Q, V.

*(PARMELIA) MUNDATA var. PULVERATA Nyl. ex Cromb., J. Linn. Soc. London, Bot. 17: 395. 1879. V. P. physodes var. pulverata Muell.-Arg., Flora 66: 176. 1883. Bibby & Smith (1954); Muell.-Arg. (1892a); Shirley (1893a). Q, V, WA.

*(PARMELIA) MUNDATA f. SOREDIOSA Bitt., Hedwigia 11: 255. 1901. V.

H. PHYSODES (L.) Nyl., Lich. Envir. Paris p. 39. 1896. Lichen physodes L., Sp. Pl. p. 114. 1753. Parmelia physodes Ach., Method. Lich. p. 250. 1803. Krempelhuber (1868); F. Mueller (1881); Shirley (1892b); Stirton (1899b); Willis (1953). NSW, Q, V, WA.

(PARMELIA) PHYSODES var. LABROSA Ach., Lichenogr. Univ. p. 493. 1810. Hellbom (1896). Sine loc.

*(PARMELIA) PHYSODES var. LEUCINA Muell.-Arg., Bull. Herb. Boiss. 4: 90. 1896. Muell.-Arg. (1898). "Dargo."

*(PARMELIA) PHYSODES var. MESOTROPA Muell.-Arg., Flora 69: 257. 1886. NSW.

(PARMELIA) PHYSODES var. OBSCURATA Ach., Syn. Lich. p. 218. 1814. P. obscurascens A. Zahlbr., Cat. Lich. Univ. 6: 35. 1929. Wilson (1887). V.

*H. PHYSODES f. SUBLUGUBRIS (Muell.-Arg.) Raes., Ann. Soc. Zool.-Bot. Fenn. "Vanamo" 18(1); 14. 1943. Parmelia physodes var. sublugubris Muell.-Arg., Flora 66: 75. 1883. Wilson (1887). V.

*(PARMELIA) PHYSODES var. RUGOSA Muell.-Arg., Flora 66: 75. 1883. V.

*(PARMELIA) PHYSODES var. SOLUTA Muell.-Arg., l.c. p. 76. NSW, V.

*(PARMELIA) PHYSODES var. TENUIS Muell.-Arg., ibid. NSW, V.

*(PARMELIA) PULCHRILOBATA Bitter, Hedwigia 11: 244. 1901. V.

H. SUBOBSCURA (Vain.) Poelt, Mitt. Bot. Staatssamml. Muenchen 4: 297-299. 1962. Parmelia subobscura Vain., Ark. Bot. 8: 33. 1909. Filson (1970) makes a superfluous new combination, no records mentioned.

*H. SUBPHYSODES (Kremp.) R. Filson, Victoria Naturalist 87: 325. 1970. Parmelia subphysodes Kremp., Verh. zool.-bot. Ges. Wien 30: 338. 1880. V. Bitter (1901); Frey (1967); Muell.-Arg. (1887c, equated with P. physodes var. pulverata Muell.-Arg.).

*(PARMELIA) SUBTERES Bitter, Hedwigia 40: 265. 1901. Q.

*(PARMELIA) TUBULARIS Tayl., Phytologist 1: 1096. 1844. NSW.

H. VITTATA (Ach.) Gasilien, Act. Soc. Linn. Bordeaux 53: 66. 1898. Parmelia physodes var. vittata Ach., Method. Lich. 250. 1803. P. vittata Roehl., Deutschl. Fl. 3(2): 109. 1813. Wilson (1887). V.

LAURERA

*L. CHRYSOCARPA (Muell.-Arg.) A. Zahlbr., Cat. Lich. Univ. 1: 503. 1922. Bathelium chrysocarpum Muell.-Arg., Hedwigia 30: 54. 1891. Bailey (1891); Letrouit-Galinou (1957); Muell.-Arg. (1895b); Shirley (1891b). Q.

L. CUMINGII (Mont.) A. Zahlbr., Cat. Lich. Univ. 1: 503. 1922. Trypethelium cumingii Mont. ex Hook., London J. Bot. 4: 5. 1845. Letrouit-Galinou (1957); Stirton (1875). Sine loc.

L. MADREPORIFORMIS (Eschw.) Riddle var. OBSCURIOR (Bab. ex Hook.) A. Zahlbr., Cat. Lich. Univ. 1: 504. 1922. Trypethelium madreporiforme var. obscurius Bab. ex Hook., Fl. Nov. Zel. 2: 305. 1855. Letrouit-Galinou (1957). Sine loc.

L. MEGASPERMA (Mont.) A. Zahlbr. var. TASMANICA (Jatta) A. Zahlbr., Cat. Lich. Univ. 1: 505. 1922. Bathelium megasperma var. tasmanicum Jatta, Bull. Soc. Bot. Ital. p. 259. 1911. Letrouit-Galinou (1957). Sine loc.

LECANACTIS

L. CONIOCHLORA (Mont. & v. d. Bosch in Mont.) A. Zahlbr., Cat. Lich. Univ. 2: 534. 1923. Lecidea coniochlora Mont. & v. d. Bosch in Mont., Syll. Gen. Sp. Crypt. p. 342. 1856. Bailey (1881, 1883). Q.

*L. MICROCARPELLA (Muell.-Arg.) A. Zahlbr., Cat. Lich. Univ. 2: 539. 1923. Opegrapha microcarpella Muell.-Arg., Bull. Herb. Boiss. 3: 318. 1895. Shirley (1896). Q.

*L. PLATYGRAPHOIDES (Muell.-Arg.) A. Zahlbr., Cat. Lich. Univ. 2: 541. 1923. Opegrapha platygraphoides Muell.-Arg., Hedwigia 32: 132. 1893. Shirley (1893a). Q.

L.QUASSIAE (Fée) A. Zahlbr., Cat. Lich. Univ. 2: 544. 1923. Lecidea quassiae Fée, Suppl. Essai Crypt. Ecorc. Officin. p. 104. 1837. *Opegrapha plurilocularis var. pruinosa Muell.-Arg., Flora 65: 504. 1882. Opegrapha quassiae Muell.-Arg., Mem. Soc. Phys. Hist. Nat. Geneve 29(8): 19. 1887. Muell.-Arg. (1891a). Q.

*L. QUASSIAE var. OBFUSCATA (Muell.-Arg.) A. Zahlbr., Cat. Lich. Univ. 2: 545. 1923. Opegrapha plurilocularis var. obfuscata Muell.-Arg., Flora 65: 505. 1882. O. quassiae var. obfuscata Muell.-Arg., Rev. Mycol. 9: 86. 1887. Q.

(Lecanactis)

L. QUASSIAE var. PLURILOCULARIS (Nyl.) A. Zahlbr., Cat. Lich. Univ. 2: 545. 1923. Lecidea premnea var. plurilocularis Nyl., Ann. Sci. Nat., Bot. (4) 15: 49. 1861. Opegrapha plurilocularis Muell.-Arg., Flora 65: 331. 1882. Q.

LECANIA

*L. CHLARONOIDES Muell.-Arg., Bull. Herb. Boiss. 3: 634. 1896. NSW.

L. CYRTELLA (Ach.) Th. Fr., Lichenogr. Scand. p. 294. 1871. Lecidea cyrtella Ach., Method. Lich. p. 67. 1803. Patellaria cyrtella Muell.-Arg., Bull. Herb. Boiss. 1: 47. 1893. Zahlbruckner (Cat. Lich. Univ. 5: 726. 1928) equates this with L. erysibe, a saxicolous species, but the habitat, on bark of Sambucus nigra, suggests otherwise.

*L. FABACEA (Muell.-Arg.) Muell.-Arg., Bull. Herb. Boiss. 3: 634. 1895. Lecanora fabacea Muell.-Arg., Hedwigia 32: 124. 1893 (basionym overlooked by Zahlbruckner, Cat. Lich. Univ. 5: 731. 1928). V.

*L. MOLLIUSCULA Muell.-Arg., Hedwigia 34: 29. 1895. V.

*L. MUELLERIANA A. Zahlbr., Cat. Lich. Univ. 5: 736. 1928. L. subsquamosa Muell.-Arg., Bull. Herb. Boiss. 3: 634. 1895 (non Muell.-Arg., 1889). V.

*L. NODULOSA (Stirt.) A. Zahlbr., Cat. Lich. Univ. 5: 736. 1928. Lecidea nodulosa Stirt. ex Bailey, Queensland Agric. J. 5: 488. 1879 (incorrectly given as Lecanora nodulosa by Zahlbr., Cat. Lich. Univ. 5: 736. 1928). Lecidea nodulosa Stirt. is a later homonym (non Koerb., 1861). Q.

*L. SANGUINOLENTA (Stirt. ex Bailey) A. Zahlbr., Cat. Lich. Univ. 5: 742. 1928. Lecidea sanguinolenta Stirt. ex Bailey, Queensland Agr. J. 5: 488. 1899 (not Lecanora sanguinolenta as given erroneously by Zahlbr., Cat. Lich. Univ. 5: 742. 1928). Q. Except for the septate spores, Stirton's description could apply to the common Lecidea russula. Zahlbruckner possibly put it into Lecania because of spore septation. It is quite likely that Stirton was intending to make the identification Callopisma sanguinolentum, which is synonymous with L. russula.

*L. SUBSQUAMOSA Muell.-Arg. var. LEPROSA (Muell.-Arg.) A. Zahlbr., Cat. Lich. Univ. 5: 745. 1928. Lecania selenospora var. leprosa Muell.-Arg., Bull. Herb. Boiss. 3: 634. 1895 (not selenispora as given by Zahlbr., Cat. Lich. Univ. 5: 745. 1928). Muell.-Arg. (1893). V.

LECANORA

*L. ALBELLARIA Muell.-Arg., Bull. Herb. Boiss. 3: 632. 1895. Shirley (1896, 1898). Q.

*L. ALLIGATA Stirt. ex Bailey, Queensland Agr. J. 5: 38. 1899. Q.

L. ALLOPHANA (Ach.) Roehl., Deutschl. Flora 3(2): 82. 1813. L. subfusca var. allophana Ach., Lichenogr. Univ. p. 395. 1810. Muell.-Arg. (1891a); Shirley (1891a); Wilson (1888). Q, V.

L. ATRA (Huds.) Ach., Lichenogr. Univ. p. 344. 1810. Lichen ater Huds., Fl. Angl. 1: 445. 1762. Parmelia atra Wibel, Primit. Fl. Werthem. p. 325. 1799. Fries (1846-47); Knight (1882); F. Mueller (1881); Shirley (1889a); Wilson (1887, 1889c). NSW, Q, V, WA.

*L. ATRA var. AUSTRALICA Raes., Arch. Soc. Zool. Bot. Fenn. "Vanamo" 3: 179. 1949. Q.

*L. ATRA var. IMMARGINATA Knight ex Bailey, Syn. Pl. Queensland Suppl. p. 72. 1886. Shirley (1889a). Q.

*L. ATRA var. SERIALIS Muell.-Arg., Bull. Herb. Boiss. 3: 632. 1895. Shirley (1896). Q.

L. ATRA var. VIRENS Muell.-Arg., Flora 65: 484. 1882. Shirley (1893a). Q.

L. BADIA (Hoffm.) Ach., Lichenogr. Univ. p. 407. 1810. Verrucaria badia Hoffm., Flecht. Deutschl. p. 182. 1796. NSW: Mt. Kosciusko, 6500 ft., 5 March 1968, Weber & McVean L-49753; 10 Nov. 1967, Weber & McVean L-47235. ACT: Brindabella Mts., 14 Dec. 1967, Weber L-46715 (COLO).

*L. BRISBANENSIS A. Zahlbr., Cat. Lich. Univ. 5: 400. 1928. L. subimmersa Muell.-Arg., Hedwigia 32: 124. 1893 (non Vain., 1890). Shirley (1893a). Q.

L. CAESIORUBELLA Ach. ssp. GLAUCOMODES (Nyl.) Imshaug & Brodo, Nova Hedwigia 12: 15. 1966. L. glaucomodes Nyl., Flora 59: 509. 1876. L. australiensis A. Zahlbr. Cat. Lich. Univ. 5: 393. 1928 (nom. nov. for *L. leucoma Nyl., Flora 47: 268. 1864, non Roehl., 1813). Bibby & Smith (1954); Fries (1846-47); Hue (1890-92); Imshaug & Brodo (1966); F. Mueller (1881); Muell.-Arg. (1891a, 1892b, 1893); Shirley (1891a, 1892b). Q, V, WA.

L. CAMPESTRIS (Schaer.) Hue, Bull. Soc. Bot. France 35: 47. 1888. Parmelia subfusca var. campestris Schaer., Lich. Helvet. Spicil., Sect. 8: 391. 1839. Muell.-Arg. (1892b). WA.

L. CARNEOLUTESCENS Nyl., Flora 61: 380. 1858. Muell.-Arg. (1893). V.

L. CARPINEA (L.) Vain., Medd. Soc. Faun. Fl. Fenn. 14: 23. 1888. Lichen carpineus L., Sp. Pl. p. 1141. 1753. Lecanora angulosa Ach., Lichenogr. Univ. p. 364. 1810. Knight (1882). NSW.

L. CHLARONA (Ach.) Nyl., Flora 65: 250. 1872. L. distincta var. chlarona Ach., Lichenogr. Univ. p. 397. 1810. Bailey (1881); Hue (1890-92); Nylander (1886); Shirley (1894a). Q.

L. CHLAROTERA Nyl., Bull. Soc. Linn. Normand. (2) 6: 274. 1872. L. subfusca var. chlarotera Harm., Bull. Soc. Bot. France 56: 86. 1909. Stirton (1899b). Q.

L. COILOCARPA (Ach.) Nyl. ex Norrlin in Not. Saellsk. Faun. Fl. Fenn. Foerhandl. 13: 330. 1871-74. L. subfusca var. coilocarpa Ach., Lichenogr. Univ. p. 393. 1810. Shirley (1889a). Q.

L. CONIZAEA (Ach.) Nyl., Flora 55: 249. 1872. L. expallens var. conizaea Ach., Lichenogr. Univ. p. 374. 1810. Hue (1890-92); Muell.-Arg. (1891a); Shirley (1889a). Q.

*L. CONNIVENS Muell.-Arg., Nuovo Giorn. Bot. Ital. 23: 289. 1891. Shirley (1891b). Q.

*L. CONTINUA Knight ex Shirley, Proc. Roy. Soc. Queensland 6: 134. 1889. Q.

*L. CONVEXELLA Raes., Arch. Soc. Zool. Bot. "Vanamo" 3: 180. 1949. NSW.

*L. (?) CORYSTA Knight, Trans. Linn. Soc. London, Bot. 2: 47. 1882. NSW.

L. CRENULATA (Dicks.) Hook. in Sm., Engl. Flora 5: 194. 1844. Lichen crenulatus Dicks., Fasc. Pl. Crypt. Brit. 3: 14. 1793. Lecanora caesioalba Koerb., Parerg. Lich. p. 82. 1859. Muell.-Arg. (1893). V.

(Lecanora)

L. DISPERSA (Pers.) Sommerf., Suppl. Fl. Lappon. p. 96. 1826. Lichen dispersus Pers., Ann. Bot. (Usteri) 1: 27. 1794. Lecanora flotowiana Spreng., Neue Entdeck. 1: 221. 1820. L. galactina var. dispersa Ach., Lichenogr. Univ. p. 424. 1810. Muell.-Arg. (1893a); Shirley (1889e, 1891a). Q, V.

*L. ELATINOIDES Raes., Arch. Soc. Zool. Bot. "Vanamo" 3: 179. 1949. NSW.

*L. ELATINOIDES var. STRAMINEA Raes., loc. cit., p. 180. NSW.

L. EPIBRYON Ach., Lichenogr. Suec. Prodrom. p. 79. 1798. L. subfusca var. bryontha Hazsl., Magy. Birod. Zuzmo-Flor. p. 113. 1884. Muell.-Arg. (1893). V.

L. FIBROSA Muell.-Arg., Flora 71: 140. 1888. Muell.-Arg. (1893). V.

*L. FLAVIDOFUSCA Muell.-Arg., Bull. Herb. Boiss. 3: 633. 1895. Shirley (1896). Q.

*L. FLAVOSTRAMINEA (Muell.-Arg.) A. Zahlbr., Cat. Lich. Univ. 5: 621. 1928. Placodium flavostramineum Muell.-Arg., Hedwigia 34: 29. 1895. V.

L. FRUSTULOSA (Dicks.) Ach., Lichenogr. Univ. p. 405. 1810. Lichen frustulosus Dicks., Fasc. Pl. Crypt. Brit. 3: 13. 1793. Muell.-Arg. (1893a). V.

L. GANGALEOIDES Nyl., Flora 55: 354. 1872. Hue (1890-92); Nylander (1886). NSW.

*L. GLAUCOFLAVENS Muell.-Arg., Bull. Herb. Boiss. 1: 39. 1893. V.

*L. GLAUCOLIVIDA (Muell.-Arg.) A. Zahlbr., Cat. Lich. Univ. 5: 624. 1928. Placodium glaucolividum Muell.-Arg., Nuovo Giorn. Bot. Ital. 23: 388. 1891. Shirley (1891b). Q.

*L. GLEBULARIS (Muell.-Arg.) A. Zahlbr., Cat. Lich. Univ. 5: 624. 1928. Placodium glebulare Muell.-Arg., Flora 71: 204. 1888. SA.

*L. HYALINESCENS Muell.-Arg., Flora 65: 484. 1882. NSW.

*L. IMPERFECTA (Muell.-Arg.) A. Zahlbr., Cat. Lich. Univ. 5: 625. 1928. Placodium imperfectum Muell.-Arg., Bull. Herb. Boiss. 4: 93. 1896. Muell.-Arg. (1898). Q.

*L. INTERJECTA Muell.-Arg., Nuovo Giorn. Bot. Ital. 23: 390. 1891. Bailey (1891); Shirley (1891b, 1893a). Q.

*L. KNIGHTIANA Muell.-Arg., Bull. Herb. Boiss. 3: 633. 1895. Shirley (1896). Q.

*L. LACTEOLA Muell.-Arg., op. cit. 1: 38. 1893. V. Shirley (1893a). Q, V.

*L. LAEVISSIMA Knight ex Shirley, Proc. Roy. Soc. Queensland 6: 133. 1889. Q.

*L. LINEOLATA Muell.-Arg., Bull. Herb. Boiss. 1: 38. 1893. V.

*L. MARGARODES (Koerb.) Nyl., Flora 47: 266. 1864. Zeora margarodes Koerb., Abh. Schlesisch. Ges. vaterl. Kultur 2: 31. 1862. Hue (1890-92). Sine loc.

L. POLYTROPA (Ehrh.) Rabenh., Deutschl. Kryptog.-Flora 2: 37. 1845. Verrucaria polytropa Ehrh. ex Hoffm., Deutsch. Fl. p. 196. 1796. NSW. Mount Kosciusko, 6500-7000 ft., 10 Nov. 1967, Weber & McVean L-47182, 47245, 49749 (COLO).

*L. PULVERATA Stirt., Trans. Proc. Roy. Soc. Victoria 17: 70. 1881. Bailey (1883); F. Mueller (1881); Shirley (1889a). Q.

*L. QUEENSLANDICA Knight ex Shirley in Bailey, Syn. Queensland Fl. 2 Suppl. p. 85. 1888. (as 'queenslandiae.') Shirley (1888a, 1889a, 1892b). Q.

L. RADIOSA (Hoffm.) Schaer., Enum. Critic. Lich. Europ. p. 61. 1850. Lichen radiosus Hoffm., Enum. Lich., tab. IV, fig. 5. 1784. Muell.-Arg. (1893a). V. This report very likely is a misidentification of a Placopsis species.

*L. MELACARPELLA Muell.-Arg., Bull. Herb. Boiss. 3: 633. 1895. Shirley (1896). Q.

*L. MELANOMMATA Knight ex Bailey, Syn. Queensland Fl. 1. Suppl. p. 71. 1886. Shirley (1889a). Q.

*L. MUNDULA Stirt., Trans. Proc. Roy. Soc. Victoria 17: 76. 1881. F. Mueller (1881). V.

*L. OBLUTESCENS Nyl., Lich. Nov. Zel. p. 64. 1888. Hue (1890-92). NSW.

L. PALLIDA (Schreb.) Rabenh. Deutschl. Kryptog.-Flora 2: 34. 1845. Lichen pallidus Schreb., Spicil. Flor. Lipsiens. p. 133. 1771. Lecanora albella (Pers.) Ach., Lichenogr. Univ. p. 369. 1810. Shirley (1889a). Q. Probably a misdetermination of L. caesiorubella.

*L. PERMINUTA Muell.-Arg., Bull. Herb. Boiss. 1: 39. 1893. V.

*L. PHAEOCARPA Muell.-Arg., Flora 70: 321. 1887. NSW.

L. PINGUIS Tuck., Proc. Amer. Acad. Arts Sci. 6: 268. 1866. Bailey (1881, 1883); F. Mueller (1881, 1883); Shirley (1889a). Q. Reported as corticolous! L. pinguis is a saxicolous endemic of the American Pacific coast.

*L. PLUMOSA Muell.-Arg., Flora 65: 484. 1882. Shirley (1889a). Q.

L. RHYPODERMA (Knight) Muell.-Arg., Nuovo Giorn. Bot. Ital. 23: 389. 1891. Lecidea rhypoderma Knight, Trans. New Zealand Inst. 12: 375. 1880. (Zahlbruckner, Catalogus 3: 761. 1925, lists this as synonym of Lecidea furfuracea Pers.). Shirley (1889a, 1891b). Q. Muell.-Arg. was evidently unaware of Knight's valid publication of the taxon in 1880, basing his name on Knight's, "solum nomen."

L. RUBINA (Vill.) Ach., Lichenogr. Univ. p. 412. 1810. Lichen rubinus Vill., Hist. Pl. Dauphin. 3: 977. 1789. Placodium chrysoleucum (Sm.) Link, Grundr. Kraeuterk. 3: 190. 1833. Muell.-Arg. (1893). V.

L. RUPICOLA (L.) A. Zahlbr., Cat. Lich. Univ. 5: 525. 1928. Lichen rupicolus L., Mantissa 1: 132. 1767. Lecanora sordida (Chev.) Th. Fr., Nova Acta Reg. Soc. Scient. Upsal. (3)3: 215. 1861. Shirley (1889a). Q.

*L. RUPICOLA f. OBSCURATA (Muell.-Arg.) A. Zahlbr., Cat. Lich. Univ. 5: 531. 1928. Lecanora sordida var. glaucoma f. obscurata Muell.-Arg., Bull. Herb. Boiss. 1: 37. 1893. V.

*L. RUTILESCENS Stirt. ex Bailey, Queensland Dept. Agr. J. 5: 37. 1899. Q.

*L. SOLENOSPORA Muell.-Arg., Bull. Herb. Boiss. 1: 38. 1892. V.

L. SOREDIIFERA Fée, Essai Crypt. Ecorc. Officin. p. 114. 1824. Shirley (1889a). Q. The Sayer and Hartman collections cited (Mel!) belong to Lecidea granifera.

*. SPHAEROSPORA Muell.-Arg., Hedwigia 31: 196. 1892. *L. grandinosa (Muell.-Arg.) A. Zahlbr., Cat. Lich. Univ. 5: 625. 1928. Placodium grandinosum Muell.-Arg., Bull. Herb. Boiss. 1: 34. 1893. V. Willis (1953). V. WA.

L. SUBCARNEA (Sw.) Ach., Lichenogr. Univ. p. 365. 1810. Lichen subcarneus Sw. ex Westr., Kongl. Svenska Vetensk.-Akad. Handl. p. 126. 1791. Lecanora sordida var. subcarnea Th. Fr., Nova Acta Reg. Soc. Scient. Upsal. (3) 3: 215. 1861. Muell.-Arg. (1893). V.

L. SUBFUSCA (L.) Ach., Lichenogr. Univ. p. 393. 1810. Lichen subfuscus L., Sp. Pl. p. 1142. 1753. Patellaria subfusca Wibel, Primit. Fl. Werthem. p. 324. 1799. Bailey (1883); F. Mueller (1881); Shirley (1889c,d); Turner (1905). Q.

*L. SUBFUSCA var. CIRCUMPLUMESCENS Nyl., Flora 69: 324. 1886. Hue (1890-92). NSW.

L. SUBFUSCA var. COMPACTA Muell.-Arg., Rév. Mycol. 10: 61. 1888. Muell.-Arg. (1891a); Shirley (1891a). Q.

*L. SUBFUSCA var. CONJUNGENS Muell.-Arg., Bull. Herb. Boiss. 3: 632. 1895. Shirley (1896). Q.

L. SUBFUSCA var. SUBGRANULATA Nyl., Ann. Sci. Nat., Bot. (5) 7: 310. 1867. Bailey (1891a); Muell.-Arg. (1891b); Shirley (1891b, 1893a). Q.

*L. SUBFUSCELLA Raes., Arch. Soc. Zool. Bot. "Vanamo" 3: 180. 1949. NSW.

*L. SUBPALLIDA Knight, Trans. Linn. Soc. London, Bot. 2: 48. 1882. NSW.

*L. SUBPINIPERDA Knight, ibid. NSW.

*L. SUBPURPUREA Stirt. ex Bailey, Queensland Agr. J. 5: 487. 1899. Q.

*L. SUBUMBRINA Muell.-Arg., Bull. Herb. Boiss. 3: 632. 1895. NSW.

L. UMBRINA (Ehrh.) Roehl., Deutschl. Flora 3(2): 75. 1813. Lichen umbrinus Ehrh., Pl. Crypt. Exsicc. no. 245. 1793. Knight (1882); Muell.-Arg. (1893); Shirley (1889a). NSW, Q, V.

*L. WATTII Stirt., Trans. Proc. New Zealand Inst. 30: 383. 1897 (1898). Stirton did not describe this from Australia but from a Himalayan collection. Zahlbruckner mistakenly gave its provenance as Australia. Furthermore, it is clear from the description that the plant is a Haematomma nearly related to H. puniceum.

*L. WILSONII Muell.-Arg., Bull. Herb. Boiss. 1: 39. 1893. V.

LECIDEA

*L. ABERRATA Stirt., Trans. Proc. Roy. Soc. Victoria 17: 71. 1881. Bailey (1881, 1883); F. Mueller (1881); Shirley (1889a). Q.

L. ALBOCOERULESCENS (Wulf.) Ach., Method. Lich. p. 52. 1803. Lichen albocoerulescens Wulf. in Jacq., Collect. Bot. 2: 184. 1788. Muell.-Arg. (1891a, 1893); Shirley (1893a). Q, V.

L. ANGOLENSIS Muell.-Arg., Linnaea 63: 35. 1880. Bailey (1891a); Muell.-Arg. (1891b); Shirley (1891a). Q.

L. ARMENIACA (DC. ex Lam. & DC.) Fries, Syst. Orb. Veget. 1: 286. 1825. Rhizocarpon armeniacum DC. ex Lam. & DC., Fl. Franc. ed. 3, 2: 367. 1805. NSW: Kosciusko State Park, trail to Northcote Pass, 7000 ft., 10 Nov. 1967, Weber & McVean L-47190 (COLO).

*L. ASPERA Muell.-Arg., Bull. Herb. Boiss. 1: 45. 1893. V.

*L. ASPIDULA Kremp., Verh. zool.-bot. Ges. Wien 30: 341. 1880. Bailey (1883); F. Mueller (1881); Muell.-Arg. (1887a, 1893, 1895d); Shirley (1889a, 1896); Weber 1969. Q, V.

*L. ASPIDULA var. DISPERSA Muell.-Arg., Bull. Herb. Boiss. 3: 640. 1895. Shirley (1896). Q.

*L. BACIDIOIDES Muell.-Arg., Flora 65: 486. 1882. Shirley (1889a). Q.

*L. BUELLIASTRUM Muell.-Arg., op. cit. 70: 322. 1887. NSW.

L. CINNABARINA Somm., Kongl. Svenska Vetensk.-Akad. Handl. p. 114. 1823. Darbishire (1912); Hampe (1853); Hue (1890-92); Muell.-Arg. (1893); Weber 1969. V.

L. CONFLUENS (Web.) Ach., Method. Lich. p. 40. 1803. Lichen confluens Web., Spicil. Flor. Goettingens. p. 180. 1778. Shirley (1889a). Q.

L. CONTIGUA (Hoffm.) Fries, Nov. Sched. Critic. p. 14. 1827. Verrucaria contigua Hoffm., Deutschl. Fl. p. 184. 1796. Bailey (1881, 1883); F. Mueller (1881); Muell.-Arg. (1891a, 1893); Shirley (1889a). Q, V.

L. CONTIGUA var. FLAVICUNDA (Ach.) Nyl., Flora 38: 675. 1855. Lecidea flavicunda Ach., Lichenogr. Univ. p. 166. 1810. Muell.-Arg. (1893). V.

L. CONTIGUA f. LEPROSA Leight., Lich. Fl. Great Brit. p. 293. 1871. Muell.-Arg. (1893). V.

*L. CONTIGUA var. UMBONIFERA Muell.-Arg., Bull. Herb. Boiss. 1: 46. 1893. V.

L. CONVEXA (Fr.) Th. Fr. var. HYDROPHILA (Fr.) Th. Fr., Lichenogr. Scand. 1: 507. 1874. Lecidea hydrophila Fr., Kgl. Sv. Vetensk.-Akad. Nya Handl. p. 256. 1822. L. contigua var. hydrophila Fr., Nov. Sched. Critic. p. 16. 1827.

*L. CRASSILABRA Muell.-Arg., Hedwigia 32: 127. 1893. V.

*L. CRYSTALLIFERA Tayl., London J. Bot. 6: 148. 1847. WA. Psora crystallifera Muell.-Arg., Flora 71: 140. 1888. P. testudinea Muell.-Arg., l.c. p. 139. P. psammophila Muell.-Arg., Hedwigia 31: 194. 1892. Lecidea psammophila A. Zahlbr., Cat. Lich. Univ. 3: 889. 1925. Hue (1890-92); Muell.-Arg. (1888d, 1893c). WA.

L. DECIPIENS (Hoffm.) Ach., Method. Lich. p. 80. 1803. Psora decipiens Hoffm., Descr. Adumbr. Pl. Lich. 2: 68. 1794. Biatora decipiens Fr., Lichenogr. Europ. Reform. p. 252. 1831. *Lecidea elegans A. Zahlbr., Cat. Lich. Univ. 3: 873. 1925. Psora elegans Muell.-Arg., Flora 64: 87. 1881. SA. Bibby & Smith (1954); Fries (1846-47); Hue (1890-92); F. Mueller (1881); Muell.-Arg. (1892b, 1893); Willis (1959b). V, WA.

L. DECIPIENS var. ALBOMARGINATA (Muell.-Arg.) A. Zahlbr., Cat. Lich. Univ. 3: 871. 1925. Psora decipiens var. albomarginata Muell.-Arg., Flora 64: 88. 1881. Muell.-Arg. (1893). V.

*L. DIAPHAENENTA Knight, Trans. Linn. Soc. London, Bot. 2: 46. 1882. NSW.

L. ELABENS Fr., Kongl. Svenska Vetensk.-Akad. Handl. p. 256. 1822. Muell.-Arg. (1893); Weber (1969). ACT, V.

*L. ENTEROPHAEA Knight, Trans. Linn. Soc. London, Bot. 2: 46. 1882. NSW.

*L. ENTEROXANTHA Knight, ibid. NSW.

L. EXIGUA Chaub. in St.-Amans, Flore Agenaise, p. 478. 1821. Shirley (1891a). Q.

*L. FERAX Muell.-Arg., Bull. Herb. Boiss. 1: 45. 1893. Maiden (1898). V.

*L. FERAX var. ATHALLINA Muell.-Arg., loc. cit., p. 46. V.

*L. FERAX var. GEOGRAPHICA Muell.-Arg., op. cit. 3: 641. 1893. Shirley (1896). Q.

L. FLEXUOSA (Fr.) Nyl., Act. Soc. Linn. Bordeaux 21: 356. 1856 (non Fries, 1825). Biatora flexuosa Fries, Nov. Sched. Critic. p. 11. 1826. Muell.-Arg. (1893). V.

L. FRIESII Ach. in Liljebl., Utkast t. Svensk Fl. p. 610. 1816. Weber (1969). ACT.

*L. FUMOSELLA Muell.-Arg., Bull. Herb. Boiss. 1: 47. 1893. V.

L. FUSCOATRA (Hoffm.) Ach., Method. Lich. p. 44. 1803. Verrucaria fuscoatra Hoffm., Deutschl. Flora p. 181. 1796. Muell.-Arg. (1893). V.

*L. GLAUCA Tayl., London J. Bot. 6: 149. 1847. WA. *Psora endochlora Muell.-Arg., Flora 71: 204. 1888 (non Lecidea endochlora Tayl., 1847). SA.

L. GRANIFERA (Ach.) Vain., Cat. Welwitsch. Afr. Pl. 2: 424. 1901. Lecanora granifera Ach., Syn. Lich. p. 164. 1814. Lecidea aurigera Fee, Essai Crypt. Ecorc. Officin. p. 106. 1824. Muell.-Arg. (1891a); Shirley (1889a, 1891a). Q.

*L. IMMARGINATA R. Br. ex Cromb., J. Linn. Soc. London, Bot. 17: 400. 1879. Hue (1890-92). NSW.

*L. INSULANA Muell.-Arg., Bull. Herb. Boiss. 3: 640. 1895. Shirley (1896). Q.

*L. INTERJUNCTA Nyl., Flora 47: 269. 1864. Hue (1890-92). Sine loc.

L. INTERVERTENS Nyl., Lich. Nov. Zel. p. 79. 1888. (= L. canorufescens Kremp., fide Zahlbr., Cat. Lich. Univ. 3: 744. 1925). V.

*L. KNIGHTIANA A. Zahlbr., Cat. Lich. Univ. 3: 604. 1925. L. microspora Knight, Trans. Linn. Soc. London, Bot. 2: 47. 1882. (non Nyl!). NSW.

*L. KORUNDENSIS Raes., Arch. Soc. Zool. Bot. Fenn. "Vanamo" 3: 183. 1949. Q.

*L. LEIOPLACA Muell.-Arg., Flora 80: 61. 1887. Shirley (1889a,b). Q.

*L. LEPTOCARPA Nyl. var. TERRIGENA Raes., Arch. Soc. Zool. Bot. Fenn. "Vanamo" 3: 183. 1949. NSW.

L. LEPTOLOMA Muell.-Arg., Flora 64: 518. 1881. Bailey (1891a); Muell.-Arg. 1891b, 1893); Shirley (1891a). Q, V.

*L. LEPTOLOMOIDES Muell.-Arg., Bull. Herb. Boiss. 1: 44. 1893. V.

*L. LEPTOPLACOIDES Raes., Arch. Soc. Zool. Bot. Fenn. "Vanamo" 3: 183. 1949. NSW.

*L. LUDIBUNDA Muell.-Arg., Flora 70: 321. 1887. Shirley (1889a). Q.

*L. LUEHMANNIANA Muell.-Arg., Hedwigia 32: 128. 1893. V.

L. MEIOSPORA (Nyl.) Nyl., Bull. Soc. Linn. Normand. (2) 4: 291. 1872. Lecidea contigua var. meiospora Nyl., Lich. Scand. p. 225. 1861. Shirley (1889a). Q.

*L. (BIATORA) MINUTULA Muell.-Arg., Bull. Herb. Boiss. 1: 44. 1893. (non Nyl., 1878). Overlooked by Zahlbruckner, Cat. Lich. Univ. V.

L. MODESTA Muell.-Arg., Flora 54: 403. 1871. Lecidea botryiza Nyl. ex Stirt., Grevillea 2: 71. 1873. Wilson (1889a). V.

*L. MULTIFLORA Tayl., London J. Bot. 6: 149. 1847. Muell.-Arg. (1888d). WA.

L. MUTABILIS Fée, Suppl. Essai Crypt. Ecorc. Officin. p. 105. 1837. Shirley (1889a). Q.

*L. NESOPHILA Muell.-Arg., Bull. Herb. Boiss. 3: 641. 1895. Shirley (1896). Q.

"L. OBOVATA Stirton." Bailey (1881, nomen nudum).

*L. OCHROLEUCA Pers. in Gaudich., Voy. Uranie, Bot. p. 193. 1826. "Bai des Chiens-Marins."

*L. PACHYPHYLLA (Muell.-Arg.) A. Zahlbr., Cat. Lich. Univ. 3: 888. 1925. Psora pachyphylla Muell.-Arg., Flora 70: 319. 1887. V.

L. PALLIDOATRA Nyl., Lich. Nov. Zel. p. 106. 1888 (overlooked by Zahlbruckner, Cat. Lich. Univ.). Muell.-Arg. (1893). V.

*L. PANTHERINA (Hoffm.) Th. Fr. f. OREINOIDES (Koerb.) A. Zahlbr., Cat. Lich. Univ. 3: 650. 1925. Aspicilia oreinoides Koerb., Abhandl. Schles. Ges. vaterl. Kultur 2: 32. 1862. Lecidea lactea f. oreinoides Nyl., Flora 47: 267. 1864. Lecidea oreinoides Hochst. ex Nyl., ibid. Hue (1890-92). NSW.

L. PARASEMA (Ach.) Ach., Method. Lich., p. 35. 1803. Lichen parasemus Ach., Lich. Suec. Prodrom. p. 64. 1798. Lecidea enteroleuca Ach., Lichenogr. Univ. p. 177. 1810. L. parasema var. enteroleuca Nyl., Act. Soc. Linn. Bordeaux 21: 370. 1856. F. Mueller (1881); Muell.-Arg. (1893); Tate (1887). SA, V, WA.

*L. PELOCHROA Nyl., Flora 47: 269. 1864. Hue (1890-92). Sine loc.

*L. PELOPHAEA Nyl., op. cit. 69: 324. 1886. Hue (1890-92). NSW.

*L. PHAEOCARPA Knight ex Shirley, Proc. Roy. Soc. Queensland 6: 171. 1889. Q.

L. PIPERIS (Spreng.) Nyl., Flora 52: 121. 1869. Lecanora piperis Spreng., Kongl. Svenska Vetensk.-Akad. Handl. p. 49. 1820. Shirley (1889a). Q.

L. PIPERIS var. MELANOCARPA Muell.-Arg., Flora 67: 689. 1884. (=L. piperis var. brasiliensis [Eschw.] A. Zahlbr., Cat. Lich. Univ. 3: 811. 1925). Shirley (1893a). Q.

*L. PLANATA Muell.-Arg., Flora 70: 61. 1887. L. plana Kremp., Verh. zool.-bot. Ges. Wien 30: 340. 1880 (non Lahm). Bibby & Smith (1954); F. Mueller (1881); Muell.-Arg. (1887a, 1892b, 1887c); Tate (1881). SA, WA.

*L. PLICATULA (Muell.-Arg.) A. Zahlbr., Cat. Lich. Univ. 3: 889. 1925. Psora plicatula Muell.-Arg., Bull. Herb. Boiss. 1: 35. 1893. V.

*L. PORPHYRIA Knight, Trans. Linn. Soc. London, Bot. 2: 46. 1882. NSW.

*L. PRUINOSULA Muell.-Arg., Flora 65: 486. 1882. Muell.-Arg. (1893). NSW, V.

*L. RAMOSELLA Raes., Arch. Soc. Zool.-Bot. Fenn. "Vanamo" 3: 182. 1949. Q.

L. RUSSULA Ach., Method. Lich., p. 61. 1803. Callopisma sanguineolentum Kremp., Verh. zool.-bot. Ges. Wien 26: 449. 1876. Bailey (1881, 1883); Krempelhuber (1880); F. Mueller (1881); Muell.-Arg. (1887c, 1891a, 1893); Shirley (1889a). Q.

*L. SCABRIDA Knight ex Bailey, Syn. Queensland Fl. 1 Suppl., p. 75. 1886. Shirley (1889a). Q.

*L. SCORIGENA Muell.-Arg., Hedwigia 34: 30. 1895. V.

*L. SELENOSPORA Muell.-Arg., Bull. Herb. Boiss. 3: 640. 1895. NSW.

L. SPEIREA (Ach.) Ach., Method. Lich. p. 52. 1803. Lichen speireus Ach., Lichenogr. Suec. Prodrom. p. 59. 1798. Shirley (1889a). Q.

*L. SUBCAERULEA Stirt. ex Bailey, Queensland Agric. J. 5: 41. 1899 (overlooked by Zahlbruckner, Cat. Lich. Univ.). Q.

*L. SUBNUBILA Stirt., Trans. Proc. Roy. Soc. Victoria 17: 71. 1881. Bailey (1881, 1883); F. Mueller (1881); Shirley (1889a). Q.

*L. SUBPROMISCUA Nyl., Flora 60: 325. 1886. Hue (1890-92). NSW.

L. SUBSIMILIS Nyl., Ann. Sci. Nat., Bot. (4) 19: 347. 1863. Muell.-Arg. (1882b); Shirley (1889a). Q.

*L. TENELLA Muell.-Arg., Bull. Herb. Boiss. 1: 44. 1893. V.

*L. TRISTICULA Muell.-Arg., loc. cit. p. 46. V.

*L. TURGENS Nyl. ex Hue, Nouv. Arch. Mus. Paris (3) 3: 133. 1891. Lecidella turgescens Koerb., Abh. Schles. Ges. vaterl. Kultur 2: 34. 1862 (non Tuck.). Hue (1890-92). NSW.

L. VARIANS Ach., Syn. Lich. p. 38. 1814. Biatora varians Eschw., Syst. Lich. p. 26. 1824. Wilson (1890b). WA.

*L. WILSONII Raes., Arch. Soc. Zool. Bot. Fenn. "Vanamo" 3: 183. 1949. Q.

*L. XYLOGENA Muell.-Arg., Bull. Herb. Boiss. 1: 44. 1893. V.

LEMPHOLEMMA

*L. HYPOLASIUM A. Zahlbr., Cat. Lich. Univ. 3: 17. 1924. Collema hypolasium Stirt. ex Bailey, Queensland Agr. J. 5: 484. 1899. Q.

L. MYRIOCOCCUM (Ach.) Th. Fr., Nova Acta Reg. Soc. Scient. Upsal. (3) 3: 381. 1861. Lichen myriococcus Ach., Lichenogr. Suec. Prodrom., p. 127. 1798. Shirley (1894). V.

LEPRARIA

L. CITRINA (Schaer.) Schaer., Lich. Helv. Spicil. 4-5: 213. 1833; Lepra citrina Schaer., op. cit. 1: 2. 1823. Muell.-Arg. (1892b). WA.

L. MEMBRANACEA (Dicks.) Lett., Fedde Repert. 61: 127. 1958. Lichen membranaceus Dicks., Pl. Cryptog. Brit. 2: 21. 1790. Amphiloma lanuginosum Nyl., Act. Soc. Linn. Bordeaux 21: 315. 1856. Wilson (1889a). V.

*L. SULPHURELLA Raes., Arch. Soc. Zool. Bot. Fenn. "Vanamo" 3: 188. 1949. V.

LEPROCAULON

L. ARBUSCULA (Nyl.) Nyl. ex Hue, Nouv. Arch. Mus. Paris (3) 4: 134. 1892. Stereocaulon arbuscula Nyl., Syn. Lich. 1: 253. 1860. Shirley (1889a). Q.

L. QUISQUILIARE (Leers) Choisy, Bull. Mens. Soc. Linn. Lyon 19: 166. 1950. Stereocaulon quisquiliare (Leers) Hoffm., Deutschl. Fl. p. 130. 1796. Lichen quisquiliaris Leers, Fl. Herborn. p. 264. 1775. Stereocaulon nanum (Ach.) Ach., Method. Lich. p. 315. 1803. Lichen nanus Ach., Lichenogr. Suec. Prodrom. p. 206. 1798. Krempelhuber (1868); Shirley (1889a); Wilson (1889a). Q, V.

LEPTOGIUM

*L. ATROVIRIDE F. Wils. ex Bailey, Queensland Dept. Agr. Bull. 7: 29. 1891. Q.

*L. AUSTRALE (Hook. f. & Tayl.) Muell.-Arg., Flora 70: 268. 1887. Collema australe Hook. f. & Tayl., London J. Bot. 3: 656. 1844. Collema saturninum var. australe Hook. f. & Tayl. in Hook., Fl. Antarctica 2: 541. 1847. Wilson (1892). V.

L. AZUREUM (Sw.) Mont. in Webb, Hist. Nat. Iles Canar. 3: 129. 1840. Lichen azureus Sw. ex Ach., Lich. Suec. Prodrom. p. 137. 1798. Bailey (1881, as var. "coeruleum;" 1891a); Muell.-Arg. (1887c, 1891a,b); Shirley (1891a); Watts (1903); Wilson (1887, 1891b, 1892). Q, V.

*L. BILOCULARE F. Wils., J. Linn. Soc. London, Bot. 28: 357. 1891. Wilson (1892). V.

L. BREBISSONII Mont. in Webb, Hist. Nat. Iles Canar. 3(2): 130. 1840. L. chloromelum auctt. Austr. non Nyl. L. rugatum Wils., Proc. Roy. Soc. Victoria (2) 5: 159. 1892. Collema rugatum Hook. f. & Tayl., London J. Bot. 3: 656. 1844. Bailey (1891a); Muell.-Arg. (1891b); Shirley (1891a); Weber (1971). NSW, Q, V.

*L. BULLATULUM Muell.-Arg. ex Bailey, Queensland Dept. Agr. Bull. 13: 23. 1891. Muell.-Arg. (1891b); Shirley (1891b). Q.

L. BULLATUM (Sw.) Mont., Ann. Sci. Nat., Bot. (2) 16: 113. 1841. Lichen bullatus Sw. ex Ach. Lichenogr. Suec. Prodrom. p. 137. 1798. Krempelhuber (1880); Muell.-Arg. (1887c, correcting Krempelhuber's report to Physma byrsinum) F. Mueller (1881). NSW.

(Leptogium)

L. BURGESSII (L.) Mont. in Webb, Hist. Nat. Iles Canar. 3 (2): 129. 1840. Lichen burgessii L. ex Murr., Syst. Veget. p. 807. 1774. Wilson (1887, 1891b, 1892). V.

L. BYSSINUM (Hoffm.) Zw. ex Nyl., Act. Soc. Linn. Bordeaux 21: 270. 1856. Lichen byssinus Hoffm., Enum. Lich. p. 46. 1784. Shirley (1889a); Wilson 1889b, as L. byssinum var. azureum Wils.). NSW, Q.

*L. CAERULEUM F. Wils. ex Bailey, Queensland Dept. Agr. Bull. 7: 28. 1891. Q.

*L. CARNEOLUM F. Wils. ex Bailey, loc. cit. p. 29. Q.

L. CYANESCENS (Ach.) Koerb., Syst. Lich. German. p. 420. 1855. Collema tremelloides var. cyanescens Ach., Syn. Lich. p. 326. 1814. *Leptogium tremelloides var. isidiosum F. Wils., Proc. Roy. Soc. Victoria n.s. 5: 156. 1892. *L. tremelloides f. isidiosa Muell.-Arg., Flora 65: 292. 1882. NSW. Shirley (1893a, 1894). NSW, V.

(L. CYANESCENS var.) *L. CAESIUM Vain. var. PICHNEUM (Ach.) Vain., Ann. Acad. Scient. Fenn., ser. A, 15 (6): 40. 1921. Collema tremelloides var. pichneum Ach., Syn. Lich. p. 343. 1814. Sine loc. (Shirley (1889a). Q.

L. DACTYLINUM Tuck. ex Nyl., Syn. Lich. 1: 123. 1858. Wilson (1887, 1891b). V. Wilson (1892) considered this a misdetermination and named the specimens L. philorheuma F. Wils. (=L. marginellum).

L. DAEDALEUM (Flot.) Nyl., Mém. Soc. Scienc. Nat. Cherbourg 5: 91. 1857. Stephanophorus daedaleus Flot., Linnaea 17: 16. 1843. Leptogium phyllocarpum var. daedaleum Nyl., Syn. Lich. 1: 130. 1858. Shirley (1889a, 1892a); Wilson (1887, 1892). Q, V.

L. HYPOTRACHYNUM Muell.-Arg., Flora 64: 81. 1881. Shirley (1889a); Turner (1905); Wilson (1892). Q, V.

L. INTERMEDIUM (Arn. ex Malbr.) Arn., op. cit. 68: 212. 1885. Leptogium minutissimum var. intermedium Arn. ex Malbr., Bull. Soc. Amis Nat. Sci. Rouen 2: 363. 1866. Leptogium lacerum var. intermedium Leight., Lich. Fl. Great Brit. ed. 3: 29. 1879. Wilson (1892). V.

L. JAVANICUM (Mont. & v.d. Bosch) Mont., Syll. Gen. Sp. Crypt. p. 379. 1856. Stephanophorus javanicus Mont. & v.d. Bosch, Pl. Jungh. fasc. 4: 492. 1855. Leptogium sphinctrinum Nyl., Flora 41: 338. 1858. Wilson (1891c). Q.

L. LICHENOIDES (L.) Zahlbr., Cat. Lich. Univ. 3: 136. 1924. Wilson (1891b). V.

L. LICHENOIDES var. PULVINATUM (Hoffm.). A. Zahlbr., Cat. Lich. Univ. 3: 142. 1924. Collema pulvinatum Hoffm., Deutschl. Fl. p. 104. 1796. Leptogium lacerum var. pulvinatum Mont. in Webb, Hist. Nat. Iles Canar. 3(2): 129. 1840. Wilson (1891b, 1892). V.

L. MARGINELLUM (Sw.) S. Gray, Nat. Arr. Brit. Pl. 1: 401. 1821. Lichen marginellus Sw., Nov. Gen. Sp. Pl. p. 147. 1788. Leptogium tremelloides var. marginellum Nyl., Syn. Lich. 1: 125. 1858. Leptogium corrugatulum Nyl., l.c. p. 132. 1858. Leptogium philorheuma F. Wils., Proc. Roy Soc. Victoria, n.s. 5: 157. 1893. Bailey (1883); Krempelhuber (1880); F. Mueller (1881); Muell.-Arg. (1887c); Shirley (1888b, 1889a, 1894); Turner (1905); Wilson (1892). NSW, Q, V.

L. MENZIESII (Sm. ex Ach.) Mont., Ann. Sci. Nat., Bot. (3) 8: 223. 1852. Lichen menziesii Sm. ex Ach., Method. Lich. p. 221. 1803. Willis (1953). WA.

L. MOLUCCANUM Vain., Etud. Lich. Brésil 1: 223. 1890. Collema moluccanum Pers. in Gaudich., Voy. Uranie, Bot. p. 203. 1826. Leptogium diaphanum (Ach.) Mont., Ann. Sci. Nat., Bot. (3) 10: 134. 1848. Bailey (1881, 1883); F. Mueller (1881); Shirley (1889a). Q.

*L. PECTEN F. Wils., J. Linn. Soc. London, Bot. 28: 358. 1891. V.

L. PHYLLOCARPUM (Pers.) Mont., Ann. Sci. Nat., Bot. (3) 10: 134. 1848. Collema phyllocarpum Pers. in Gaudich., Voy. Uranie, Bot. p. 204. 1826. Bailey (1883); Krempelhuber (1880); F. Mueller (1881); Shirley (1888b, 1889d); Wilson (1889c, 1891b). NSW, Q.

L. PHYLLOCARPUM var. ISIDIOSUM Nyl., Syn. Lich. 1: 130. 1858. Bailey (1881); Bibby (1954); Muell.-Arg. (1887c, 1891a); Shirley (1889a, 1891a); Stirton (1899b). Q.

L. SINUATUM (Huds.) Mass., Mem. Lichenogr. p. 88. 1853. Lichen sinuatus Huds., Fl. Angl. ed. 2, 2: 535. 1778. Wilson (1887, 1891b, 1892). V.

L. TREMELLOIDES (L.) S. Gray, Nat. Arr. Brit. Pl. 1: 400. 1821. Lichen tremelloides L. f., Suppl. Plant. p. 450. 1781. Bailey (1883); Krempelhuber (1870, 1880); F. Mueller (1881); Muell.-Arg. (1891a); Shirley (1888b, 1889a,d, 1892b); Stirton (1899b); Turner (1905); Wilson (1891b); Zahlbruckner (1896). NSW, Q.

*L. TREMELLOIDES var. LIMBATUM (F. Wils.) Muell.-Arg., Bull. Herb. Boiss. 4: 87. 1896. Leptogium limbatum F. Wils., Victoria Naturalist 6:63. 1889. J. Linn. Soc. London, Bot. 28: 358. 1891. L. inflexum var. limbatum F. Wils., Proc. Roy. Soc. Victoria, n.s. 5: 158. 1892. Muell.-Arg. (1898). V.

*L. TREMELLOIDES var. MUSCITEGENS F. Wils., Proc. Roy. Soc. Victoria n.s. 5: 156. 1892. V.

*L. VICTORIANUM F. Wils., J. Linn. Soc. London, Bot. 28: 358. 1891. Leptogium chloromelum var. victorianum F. Wils., Victoria Naturalist 6: 62. 1889. (overlooked by Zahlbruckner, Cat. Lich. Univ. 3: 160. 1924). Wilson (1889a, 1891b, 1892). V.

*L. WILSONII A. Zahlbr., Cat. Lich. Univ. 3: 160. 1924. Leptogium denticulatum F. Wils. ex Bailey, Queensland Dept. Agr. Bull. 7: 28. 1891. (non Nyl., 1867). V.

LEPTOTREMA

*L. AEMULUM Muell.-Arg., Bull. Herb. Boiss. 3: 316. 1895. Shirley (1896). Q, V.

*L. ALBOCORONATUM Knight ex Shirley, Proc. Roy. Soc. Queensland 6: 192. 1889. Q.

*L. DIFFRACTUM Muell.-Arg., Hedwigia 30: 50. 1891. Bailey (1891a); Muell.-Arg. (1891b), Shirley (1891b). Q.

*L. FALLAX Muell.-Arg., Flora 70: 62. 1887. Muell.-Arg. (1887a); Shirley (1889a). Muell.-Arg. (1887c) states that records of T. olivaceum Mont. are misdeterminations of this. But see Ocellularia bonplandii Muell.-Arg. Q.

*L. HETEROSPORUM A. Zahlbr., Cat. Lich. Univ. 2: 635. 1923. Thelotrema heterosporum Knight ex Bailey, Syn. Queensland Flora I. Suppl., p. 72. 1886. Leptotrema mastoideum Muell.-Arg., Flora 70: 400. 1887. Muell.-Arg. (1887b); Shirley (1889a as L. "compactum"). Q.

*L. INTEGRUM Muell.-Arg., Flora 70: 399. 1887. Muell.-Arg., (1887b); Shirley (1889a,b). Q.

L. MONOSPORUM (Nyl.) Muell.-Arg., Bull. Soc. Bot. Belg. 31: 35. 1892. Thelotrema monosporum Nyl., Ann. Sci. Nat., Bot. (4) 15: 46. 1860. Shirley (1889a). Q.

*L. NITIDULUM Muell.-Arg., Bull. Herb. Boiss. 3: 315. 1895. Shirley (1896). Q.

L. PATULUM (Nyl.) Muell.-Arg., ibid. Thelotrema monosporum var. patulum Nyl., Acta Soc. Scient. Fenn. 7: 452. 1863. Shirley (1896). Q.

L. PHAEOSPORUM (Nyl.) Muell.-Arg., Flora 65: 499. 1882. Thelotrema phaeosporum Nyl., Ann. Sci. Nat., Bot. (4) 11: 242. 1859. Bailey (1881, 1883); Muell.-Arg. (1881); Shirley (1889a). Q.

*L. POLYCARPUM Muell.-Arg., Bull. Herb. Boiss. 3: 315. 1895. Shirley (1896). Q.

L. WIGHTII (Tayl.) Muell.-Arg., Flora 65: 499. 1882. Endocarpon wightii Tayl., London J. Bot. 6: 155. 1847. *Endocarpon baileyi Stirton, Trans. Proc. Roy. Soc. Victoria 17: 74. 1881. Q. Leptotrema baileyi Shirley, Proc. Roy. Soc. Queensland 6: 194. 1889. Bailey (1881, 1883); Krempelhuber (1880); Muell.-Arg. (1881, 1882b, 1891a); Shirley (1889a, 1892b, Stirton (1899b). Q.

LICHINA

L. CONFINIS (O.F. Muell.) C.A. Ag., Spec. Algarum 1: 105. 1824. Lichen confinis O.F. Muell., Icon. Pl. Daniae 5: 5. 1798. Tepper (1883); Wilson (1887, 1891b, 1892). SA, V.

L. PYGMAEA (Lightf.) C.A. Ag., Spec. Algarum 1: 105. 1821. Fucus pygmaeus Lightf., Fl. Scotica 2: 964. 1777. F. Mueller (1881); Tate (1882); Wilson (1892). NSW. SA, V. All specimens available appear to represent L. confinis.

LOBARIA

L. ADSCRIPTA Hue, Nouv. Arch. Mus. Paris (4) 2: 26. 1900. Ricasolia herbacea f. adscripta Nyl., Flora 48: 299. 1865. R. adscripta Nyl., Flora 52: 314. 1869. Stizenberger (1895), Zahlbruckner (1896). NSW.

*L. ASTICTA (Nyl.) A. Zahlbr. var. HYPOLEUCA (Muell.-Arg) A. Zahlbr., Cat. Lich. Univ. 3: 297. 1925. Ricasolia hypoleuca Muell.-Arg., Flora 65: 305. 1882. Muell.-Arg. (1883a); Shirley (1890); Stizenberger (1895). Q.

L. CRENULATA (Hook.) Trev., Lichenoth. Veneta no. 75. 1869. Parmelia crenulata Hook. ex Kunth, Syn. Pl. Aequinoct. Orb. Nov. 1: 23. 1822. Krempelhuber (1880); F. Mueller (1881); Muell.-Arg. (1887c); Shirley (1889a). Q.

L. DISCOLOR (Bory) Hue, Nouv. Arch. Mus. Paris (4) 3: 23. 1901. Sticta discolor Bory ex Del., Hist. Lich. Sticta p. 136. 1822. Cheel (1912-1914); Zahlbruckner (1896). NSW, Q.

L. DISSECTA (Sw.) Raeusch., Nomencl. Bot. p. 330. 1797. Lichen dissectus Sw., Nova Gen. et Sp. Pl. p. 147. 1788. Laurer (1827). Sine loc.

*L. HARTMANNII (Muell.-Arg.) A. Zahlbr., Cat. Lich. Univ. 3: 302. 1925. Ricasolia hartmannii Muell.-Arg., Flora 66: 45. 1883. Shirley (1889a, 1892b); Stizenberger (1895); Turner (1905). Q.

*L. LUCAEANA (Hampe) A. Zahlbr. (orth. error, "luceana"), Cat. Lich. Univ. 3: 307. 1925. Sticta lucaeana Hampe, Linnaea 17: 121. 1843. "Alpibus Novae Hollandiae."

L. PATINIFERA (Tayl.) Hue, Nouv. Arch. Mus. Paris (4) 3: 29. 1901. Parmelia patinifera Tayl. ex Hook., London J. Bot. 6: 172. 1847. Ricasolia sublaevis Nyl. ex Kremp., Flora 51: 231. 1868. NSW.

*L. PLURIMSEPTATA (Knight ex Bailey) A. Zahlbr., Cat. Lich. Univ. 3: 309. 1925. Ricasolia plurimseptata Knight ex Bailey, Syn. Queensland Fl., 2 Suppl. p. 81. 1888. Cheel (1912-1914); Shirley (1888a, 1889a, 1892b). NSW, Q.

L. PULMONARIA (L.) Hoffm., Deutschl. Flora p. 146. 1796. Lichen pulmonarius L., Sp. Pl. p. 1145. 1753. Sticta pulmonacea Ach., Lichenogr. Univ. p. 449. 1810. Lobaria pulmonacea Shirley, Proc. Roy. Soc. Queensland 6: 27. 1889. Bailey (1881, 1883); Hue (1890-92); F. Mueller (1881); Nylander (1857); Shirley (1892b); Stirton (1899b); Turner (1905); Zahlbruckner (1896). Q.

L. PULMONARIA f. HYPOMELA (Del.) Cromb., Monogr. Brit. Lich. 1: 272. 1894. Sticta pulmonacea var. hypomela Del., Hist. Lich. Sticta p. 144. 1822. Krempelhuber (1868); Shirley (1892a); Stizenberger (1895). Q.

L. PULMONARIA f. PAPILLARIS (Del.) Hue, Nov. Arch. Mus. Paris (4) 3: 31. 1901. Sticta pulmonacea var. papillaris Del., Hist. Lich. Sticta p. 144. 1822. Cheel (1912-1914); Muell.-Arg. (1891a); Shirley (1889a); Stizenberger (1895). Q, NSW.

L. QUERCIZANS (Ach.) Michx., Flora Bor.-Amer. ed. 2, p. 324. 1820. Parmelia quercizans Ach., Lichenogr. Univ. p. 464. 1810. Sticta quercizans Ach., Syn. Lich. p. 234. 1814. Ricasolia erosa Nyl., Syn. Lich. 1: 371. 1860. R. ravenelii Nyl., Acta Soc. Sci. Fenn. 7: 438. 1863. Bailey (1881, 1883); Krempelhuber (1868); F. Mueller (1881); Muell.-Arg. (1891a); Shirley (1889a, 1892b, 1893a); Stirton (1899b); Turner (1905). NSW, Q.

L. RETIGERA (Bory) Trev., Lichenoth. Veneta no. 75. 1869. Lichen retiger Bory, Voy. Quatr. Iles d'Afr. 1: 391. 1804. Bailey (1881, 1883); Cheel (1912-1914); Krempelhuber (1880); F. Mueller (1881); Muell.-Arg. (1887c); Shirley (1889a,d, 1892b); Turner (1905). NSW, Q.

L. RETIGERA f. ISIDIOSA Muell.-Arg., Flora 65: 300. 1882. Lobaria isidiosa Vain., Philipp. J. Sci. C, 8: 129. 1913. Cheel (1912-1914); Stizenberger (1895). NSW, Q.

*L. RHAPHISPORA (Knight ex Bailey) A. Zahlbr., Cat. Lich. Univ. 3: 319. 1925. Ricasolia rhaphispora Knight ex Bailey, Proc. Roy. Soc. Queensland 1: 151. 1884. Shirley (1888b, 1889a); Turner (1905). Q.

L. SCROBICULATA (Scop.) Gaertn. em. DC. in Lam. & DC., Flore Franç. ed. 3, 2: 402. 1805. Lichen scrobiculatus Scop., Fl. Carniol. ed. 2, 2: 384. 1772. Stizenberger (1895, dubiously); Zahlbruckner (1896). NSW.

*L. SPLACHNIRIMA (Hook. f. & Tayl.) A. Zahlbr., Cat. Lich. Univ. 3: 292. 1925. Parmelia splachnirima Hook. f. & Tayl., London J. Bot. 3: 645. 1844. Baeomyces squamarioides Nyl., Syn. Lich. 1: 184. 1860. Knightiella leucocarpa Muell.-Arg., Flora 69: 255. 1886. Knightiella squamarioides Muell.-Arg., op. cit. 71: 202. 1888. Wilson (1892). V.

L. STICTAEFORMIS (Schaer.) Trev., Lichenoth. Veneta no. 75. 1869. Parmelia stictaeformis Schaer. ex Moritzi, Syst. Verzeichn. p. 128. 1845-46. Ricasolia schaereri Nyl., Syn. Lich. 1: 367. 1860. Bailey (1883); Muell.-Arg. (1891a); Shirley (1889a); Stizenberger (1895). Q.

LOPADIUM

L. BIFERUM (Nyl.) A. Zahlbr., Cat. Lich. Univ. 4: 300. 1926. Lecidea bifera Nyl., Ann. Sci. Nat., Bot. (4) 15: 47. 1861. Heterothecium biferum Muell.-Arg., Flora 64: 104. 1881. Muell.-Arg. (1891a); Shirley (1892a). Q.

*L. BRISBANENSE (Knight ex Shirley) A. Zahlbr., Cat. Lich. Univ. 4: 300. 1926. Lecidea brisbaniensis Knight ex Shirley, Proc. Roy. Soc. Queensland 6: 176. 1889. Shirley (1888a). Q.

(HETEROTHECIUM) CINNABARINUM F. Wils., Victoria Naturalist 5: 31. 1888. Nomen nudum.

L. FUSCOLUTEUM (Dicks.) Mudd, Man. Brit. Lich. p. 190. 1861. Lichen fuscoluteus Dicks., Fasc. Pl. Crypt. Brit. 2: 18. 1790. Lecidea fuscolutea Ach., Kongl. Svenska Vetensk.-Akad. Nya Handl. p. 266. 1808. Heterothecium fuscoluteum Tuck., Gener. Lich. p. 176. 1872. Shirley (1889a, 1893a). Q.

L. FUSCUM Muell.-Arg., Flora 64: 108. 1881. Santesson (1952). Q.

L. LECANORELLUM Mass., Atti Imp. Regia Istit. Veneto (3) 5: 262. 1860. *Lecidea hodgkinsoniae Kremp., Verh. zool.-bot. Ges. Wien 30: 341. 1880. Muell.-Arg. (1881, 1887c). NSW.

L. LEUCOXANTHUM (Spreng.) A. Zahlbr., Sitzungsb. Kaiserl. Akad. Wiss., Math.-naturwiss. Cl. Abt. I. 111 (1): 398. 1902. Lecidea leucoxantha Spreng., Kongl. Svenska Vetensk.-Akad. Handl. p. 46. 1820 Shirley (1889a). Q.

*(SPOROPODIUM) LEUCOXANTHUM var. MICROCARPA Raes., Arch. Soc. Zool. Bot. Fenn. "Vanamo" 3: 184. 1949. Q.

L. PARABOLUM (Nyl.) A. Zahlbr., Cat. Lich. Univ. 4: 309. 1926. Lecidea parabola Nyl., Bull. Soc. Linn. Normand. (2): 90. 1868. Heterothecium parabolum Muell.-Arg., Rév. Mycol. 9: 80. 1887. Shirley (1889b). Q.

*L. PARABOLUM var. SUBVULPINUM (Muell.-Arg.) A. Zahlbr. [orthogr. error. "subalpinum"]. Cat. Lich. Univ. 4: 309. 1926. Heterothecium parabolum var. subvulpinum Muell.-Arg., Flora 70: 338. 1887, Lecidea parabola var. subvulpina Shirley, Proc. Roy. Soc. Queensland 6: 184. 1889. Q.

*L. PULCHRUM (Muell.-Arg.) A. Zahlbr., Cat. Lich. Univ. 4: 313. 1926. Heterothecium pulchrum Muell.-Arg., Nuovo Giorn. Bot. Ital. 23: 393. 1891. Bailey (1891); Shirley (1891b). Q.

L. PUIGGARII (Muell.-Arg.) A. Zahlbr., Cat. Lich. Univ. 4: 313. 1926. Heterothecium puiggarii Muell.-Arg., Flora 64: 105. 1881. Santesson (1952). NSW, Q.

*L. SAYERI (Muell.-Arg.) A. Zahlbr., Cat. Lich. Univ. 4: 313. 1926. Heterothecium sayeri Muell.-Arg., Flora 70: 338. 1887. Lecidea sayeri Shirley, Proc. Roy. Soc. Queensland 6: 184. 1889. Q.

L. VULPINUM (Tuck.) A. Zahlbr., Cat. Lich. Univ. 4: 316. 1926. Lecidea vulpina Tuck. ex Nyl., Ann. Sci. Nat., Bot. (4) 19: 354. 1863. Heterothecium vulpinum Tuck., Syn. N. Am. Lich. 2: 57. 1888. Miltidea vulpina Stirt., Trans. Proc. New Zealand Inst. 30: 386 (1897) 1898. Bailey (1881, 1883); F. Mueller (1881); Muell.-Arg. (1891a); Shirley (1889d, 1893a); Stirton (1898). Q.

*L. VULPINUM f. CORALLINUM (Muell.-Arg.) A. Zahlbr., Cat. Lich. Univ. 4: 316. 1926. Heterothecium vulpinum f. corallinum Muell.-Arg., Nuovo Giorn. Bot. Ital. 23: 393. 1891. Bailey (1891); Muell.-Arg. (1892a); Shirley (1891b). Q.

*L. VULPINUM var. GLAUCESCENS (Muell.-Arg.) A. Zahlbr., Cat. Lich. Univ. 4: 316. 1926. Heterothecium vulpinum var. glaucescens Muell.-Arg., Nuovo Giorn. Bot. Ital. 23: 393. 1891. Shirley (1892a). Q.

MARONEA

M. CONSTANS (Nyl.) Hepp, Flecht. Europ. no. 771. 1860. Lecanora constans Nyl., Mem. Soc. Imp. Scienc. Nat. Cherbourg 3: 199. 1855. Muell.-Arg. (1893). V.

MAZOSIA

M. MELANOPHTHALMA (Muell.-Arg.) R. Sant., Symb. Bot. Upsal. 12 (1): 117. 1952. Opegrapha melanophthalma Muell.-Arg., Flora 66: 348. 1883. Santesson (1952). Q.

M. PHYLLOSEMA (Nyl.) A. Zahlbr., Cat. Lich. Univ. 2: 503. 1923. Platygrapha phyllosema Nyl., Bull. Soc. Linn. Normand. (2) 7: 171. 1873. Santesson (1952). Q.

MEDUSULINA

*M. EGENELLA (Muell.-Arg.) Muell.-Arg., Bull. Herb. Boiss. 2: 93. 1894. Graphina egenella Muell.-Arg., Hedwigia 30: 52. 1891. Bailey (1891a); Shirley (1891b). Q.

M. NITIDA (Eschw.) Muell.-Arg., Bull. Herb. Boiss. 2: 92. 1894. Diorygma nitidum Eschw. in Mart., Fl. Brasil. 1: 68. 1833. Muell.-Arg. (1893). V.

MEGALOSPORA

M. MARGINIFLEXA (Hook. f. & Tayl.) A. Zahlbr., Cat. Lich. Univ. 4: 88. 1926. Lecidea marginiflexa Hook. f. & Tayl., London J. Bot. 3: 638. 1844. Patellaria marginiflexa Muell.-Arg., Flora 65: 330. 1882. Muell.-Arg. (1893). V.

*M. MELANODERMIA (Muell.-Arg.) A. Zahlbr., Cat. Lich. Univ. 4: 89. 1926. Patellaria melanodermia Muell.-Arg., Nuovo Giorn. Bot. Ital. 23: 392. 1891. Bailey (1891); Shirley (1891b). Q.

*M. RENIFORMIS (Shirley) A. Zahlbr., Cat. Lich. Univ. 4: 89. 1926. Lecidea reniformis Shirley, Proc. Roy. Soc. Queensland 6: 174. 1889. Shirley (1888a). Q.

*M. SULPHURATA Mey. & Flot., Nova Actorum Acad. Caes. Leop.-Carol. Nat. Cur. 19, Suppl. p. 228. 1843. Biatora taitensis Mont., Ann. Sci. Nat., Bot. (3) 10: 126. 1848. Lecidea taitensis Nyl., op. cit. (4) 19: 243. 1863. Patellaria sulphurata Muell.-Arg., Jahrb. Koenigl. Bot. Gart. Berlin 2: 316. 1883. Bailey (1881, 1883); F. Mueller (1881); Muell.-Arg. (1891a); Shirley (1891a). Q.

M. SULPHURATA var. CAMPYLOSPORA (Stirt.) A. Zahlbr., Cat. Lich. Univ. 4: 90. 1926. Lecidea campylospora Stirt., Trans. New Zealand Inst. 6: 238. 1873. Patellaria sulphurata var. epiglauca Muell.-Arg., Bull. Herb. Boiss. 2, appendix I, p. 64. 1894. "Patellaria taitensis var. epiglauca Nyl." (Shirley, 1892a). Q.

*M. SULPHURATA var. GALACTOCARPA (A. Zahlbr.) A. Zahlbr., Cat. Lich. Univ. 4: 90. 1926. Psorothecium taitense var. galactocarpum A. Zahlbr., Ann. Mycol. 2: 270. 1904. NSW.

M. VERSICOLOR (Fee) A. Zahlbr. in Engler & Prantl, Nat. Pflanzenfam. I. Teil, Abt. I*, p. 134. 1905. Lecanora versicolor Fée, Essai Cryptog. Ecorc. Officin. p. 115. 1824. Lecidea versicolor Nyl., Acta Soc. Scient. Fenn. 7: 461. 1863. Crombie (1880). NSW.

MELANOTHECA

M. ACHARIANA (Spreng.) Fée, Suppl. Essai Crypt. Ecorc. Officin. p. 71. 1837. Mycoporum acharii Spreng., Syst. Veg. 4(1): 242. 1827. Muell.-Arg. (1895b). Q.

*M. CINNABARINA (Knight ex Bailey) Shirley, Lich. Fl. Queensland 4: 167. 1890. Trypethelium cinnabarinum Knight ex Bailey, Syn. Queensland Fl. 1. Suppl. p. 76. 1886. Muell.-Arg. (1895b). Q.

M. CRUENTA (Mont.) Muell.-Arg., Bot. Jahrb. 6: 397. 1885. Trypethelium cruentum Mont., Ann. Sci. Nat., Bot. (2) 8: 537. 1837. *Trypethelium rubrum Knight ex Bailey, Proc. Roy. Soc. Queensland 1: 152. 1884. *Melanotheca rubra Shirley, Lich. Fl. Queensland 4: 167. 1890. Bailey (1881, 1883, 1891a); Crombie (1880); Harmand (1911); Knight (1889c); F. Mueller (1881); Muell.-Arg. (1891b, 1895b); Shirley (1895). NSW, Q,

*M. OXYSPORA Muell.-Arg., Rep. Austral. Assoc. Adv. Sci. p. 462. 1895. Shirley (1895). Q.

*M. RUBESCENS (Knight ex Shirley) Shirley, Lich. Fl. Queensland 4: 167. 1890. Trypethelium rubescens Knight ex Shirley, loc. cit. Muell.-Arg. (1895b); Shirley (1888b). Q.

M. SUBSIMPLEX Muell.-Arg., Hedwigia 30: 55. 1891. Bailey (1891); Muell.-Arg. (1891b, 1895b); Shirley (1891b). Q.

MELASPILEA

*M. ASTERISCUS (Muell.-Arg.) Muell.-Arg., Mem. Soc. Phys. et Hist. Nat. Geneve 29(8): 20. 1887. Melanographa asteriscus Muell.-Arg., Flora 65: 519. 1882. NSW. Graphis parmeliarum Knight ex Bailey, Synops. Queensland Fl. 2. Suppl. p. 88. 1888. Q. Shirley (1888a, 1891a). NSW, Q.

*M. CONGREGANS Muell.-Arg., Nuovo Giorn. Bot. Ital. 23: 397. 1891. Bailey (1891); Shirley (1891b). Q.

*M. CONGREGANTULA Muell.-Arg., Bull. Herb. Boiss. 3: 317. 1895. Shirley (1896). Q.

M. GEMELLA (Eschw.) Nyl., Ann. Sci. Nat., Bot. (5) 7: 344. 1867. Graphis scaphella var. gemella Eschw. in Mart., Fl. Brasil. 1: 88. 1833. Muell.-Arg. (1893). V.

*M. LEUCINA (Muell.-Arg.) Muell.-Arg., Mém. Soc. Phys. Hist. Nat. Genève 29(8): 19. 1887. Melanographa leucina Muell.-Arg., Flora 65: 516. 1882. Q.

*M. MICROCARPA (Muell.-Arg.) Muell.-Arg., Mém. Soc. Phys. Hist. Nat. Genève 29(8): 19. 1887. Melanographa microcarpa Muell.-Arg., Flora 65: 516. 1882. NSW.

M. OPEGRAPHOIDES Nyl., Acta Soc. Scient. Fenn. 7: 487. 1863. Shirley (1889a). Q.

(Melaspilea)

*M. STELLARIS Muell.-Arg., Bull. Herb. Boiss. 3: 317. 1895. Shirley (1896). Q.

MENEGAZZIA

*M. AENEOFUSCA (Muell.-Arg.) R. Sant., Ark. Bot. 30A (11): 13. 1942. Parmelia aeneofusca Muell.-Arg., Flora 66: 77. 1883. V. Weber (1969). NSW, V.

M. CINCINNATA (Ach.) Bitt., Hedwigia 40: 172. 1901. Parmelia cincinnata Ach., Method. Lich., p. 252. 1803. Krempelhuber (1880); F. Mueller (1881); Muell.-Arg. (1887c). NSW. V.

*M. ENTEROXANTHA (Muell.-Arg.) R. Sant., Ark. Bot. 30A (11): 12. 1943. Parmelia enteroxantha Muell.-Arg., Bull. Herb. Boiss. 4: 90. 1896. Muell.-Arg. (1898). NSW.

(M. PERTRANSITA [Stirt.] R. Sant.). *Parmelia pertransita var. phaeocarpa Muell.-Arg., Bull. Herb. Boiss. 4: 91. 1896. Muell.-Arg. (1898). Q.

M. PERTUSA (Schrank) Stein in Cohn, Krypt.-Fl. Schles. 2(2): 78. 1879. Lichen pertusus Schrank, Baierisch. Fl. 2: 519. 1789. Parmelia diatrypa Ach., Method. Lich. p. 251. 1803. Bibby & Smith (1954); Nylander (1857); F. Mueller (1881); Wilson (1887, 1889c, 1890b). V, WA.

*M. PLATYTREMA (Muell.-Arg.) R. Sant., Ark. Bot. 30A (11): 13. 1942. Parmelia platytrema Muell.-Arg., Flora 70: 60. 1887. V. Vainio (1900).

MICROTHELIA

*M. ALBA Muell.-Arg., Rep. Austral. Assoc. Adv. Sci. p. 455. 1895. [overlooked by Zahlbruckner, Cat. Lich. Univ.] Shirley (1895). Q.

M. (?) ANALTIZA (Stirt.) A. Zahlbr., Cat. Lich. Univ. 1: 255. 1921. Verrucaria analtiza Stirt., Trans. Glasgow Soc. Field Nat. 4: 95. 1876. NSW.

M. ATERRIMA (anzi) A. Zahlbr., Cat. Lich. Univ. 1: 255. 1921. Rinodina aterrima Anzi, Comment. Soc. Crittogam. Ital. 2(1): 11. 1864. NSW: on low outcrops of silcrete in a field a few miles from Adaminaby on road to Cooma, 17 Oct. 1967, Weber L-49215 (COLO).

*M. BRISBANENSIS Muell.-Arg., Rep. Austral. Soc. Adv. Sci. p. 455. 1895. Shirley (1895). Q.

M. MICULIFORMIS Muell.-Arg., Bot. Jahrb. 6: 417. 1885. Muell.-Arg. (1891a, 1895b); Shirley (1891a). Q.

*M. OBOVATA (Stirt.) Muell.-Arg., Rep. Austral. Assoc. Adv. Sci. p. 454. 1895. Verrucaria obovata Stirt., Trans. Proc. Roy. Soc. Victoria 17: 74. 1881. Pyrenula obovata Shirley, Lich. Fl. Queensland 4: 176. 1890. Bailey (1881, 1883); F. Mueller (1881); Shirley (1895). Q.

*M. QUEENSLANDICA Muell.-Arg., Rep. Austral. Assoc. Adv. Sci. p. 455. 1895. Shirley (1895). Q.

*M. SHIRLEYANA Muell.-Arg., ibid. Shirley (1895). Q.

*M. SUBGREGANS Muell.-Arg., ibid. Shirley (1895). Q.

*M. STICTARIA Muell.-Arg., Bull. Herb. Boiss. 3: 326. 1895. V.

MYCOCALICIUM

*M. AUSTRALICUM Raes., Arch. Soc. Zool.-Bot. Fenn. "Vanamo" 3: 188. 1949. NSW.

*M. OCEANICUM Raes., ibid. NSW.

MYCOPORELLUM

*M. MICROSPERMUM Muell.-Arg., Bull. Herb. Boiss. 3: 325. 1895. V.

*M. PEREXIGUUM Muell.-Arg., Nuovo Giorn. Bot. Ital. 23: 399. 1891. Bailey (1891); Shirley (1891b). Q.

MYCOPOROPSIS

*M. SORENOCARPA (Knight) Muell.-Arg., Flora 68: 515. 1885. Mycoporum sorenocarpum Knight, Trans. Linn. Soc. London, Bot. 2: 40. 1882. NSW.

NEOPHYLLIS

*N. MELACARPA (F. Wils.) F. Wils., J. Linn. Soc. London, Bot. 28: 372. 1891. Phyllis melacarpa F. Wils., Victoria Naturalist 6: 68. 1891. V. Phyllopsora melanocarpa Muell.-Arg., Hedwigia 34: 28. 1895. Lecidea dactylophylla A. Zahlbr., Cat. Lich. Univ. 3: 867. 1925. Psora dactylophylla Muell.-Arg., Bull. Herb. Boiss. 1: 35. 1893. Shirley (1894); Weber (1969). NSW, V.

NEPHROMA

N. ANTARCTICUM (Wulf.) Nyl., Syn. Lich. 1: 317. 1860. Lichen antarcticus Wulf. in Jacq., Misc. Bot. Austral. 2: 370. 1781. Krempelhuber (1880); Muell.-Arg. (1887c). Krempelhuber doubted that the specimen reported actually came from Australia! Muell.-Arg. is referring to the same report.

N. CELLULOSUM (Sm. ex Ach.) Ach., Lichenogr. Univ. p. 523. 1810. Lichen cellulosus Sm. ex Ach., Method. Lich. p. 289. 1803. Laurer (1827); Weber (1969); Wilson (1887). V.

N. HELVETICUM Ach. var. RUFUM (Bab.) Murray, Trans. Roy. Soc. New Zealand 88: 386. 1960. N. resupinatum var. rufa Bab. in Hook., Fl. New Zealand 2: 272. 1855. Murray (1960c). V.

N. LAEVIGATUM Ach., Syn. Lich. p. 242. 1814. Nephromium laevigatum Nyl., Mém. Soc. Science. Nat. Cherbourg 5: 101. 1857. Bailey (1881, 1883); Muell.-Arg. (1881); Shirley (1889a). Q. The specimens seen belong to N. tropicum.

*N. TROPICUM (Muell.-Arg.) A. Zahlbr., Cat. Lich. Univ. 3: 442. 1925. Nephromium tropicum Muell.-Arg., Flora 66: 21. 1883. Q.

NEUROPOGON

N. MELAXANTHUS (Ach.) Nyl., Syn. Lich. 1: 272. 1860. Usnea melaxantha Ach., Method Lich. p. 307. 1803. Darbishire (1912); Wilson (1888, 1890a). V. Motyka does not allow this species in Australia, but lists no other for the continent.

NORMANDINA

N. PULCHELLA (Borr.) Nyl., Ann. Sci. Nat. Bot. (4) 15: 382. 1861. Verrucaria pulchella Borr. ex Hook. & Sowerby, Suppl. Engl. Bot. 1:

tab. 2602. 1831. Normandina jungermanniae Nyl., Mém. Soc. Imp. Sci. Nat. Cherbourg 3: 191. 1855. Muell.-Arg. (1893a); Tate (1882); Weber (1969). NSW, SA, V.

OCELLULARIA

O. ALBA (Fee.) Muell.-Arg., Mém. Soc. Phys. Hist. Nat. Genève 29 (8): 6. 1887. Myriotrema album Fée, Essai Cryptog. Ecorc. Officin. p. 104. 1824. Weber (1971). Q.

*O. ALBA var. CAESIASCENS Raes., Arch. Soc. Zool. Bot. Fenn. "Vanamo" 3: 185. 1949. Q.

*O. ANNULOSA Muell.-Arg., Bull. Herb. Boiss. 3: 314. 1895. Shirley (1896). Q.

*O. BAILEYI Muell.-Arg., Hedwigia 30: 51. 1891. Bailey (1891a); Shirley (1891b). Q.

O. BONPLANDII (Fée) Muell.-Arg., Flora 65: 499. 1882. Thelotrema bonplandii Fée, Essai Cryptog. Ecorc. Officin. p. 94. 1824. T. olivaceum Mont. ex Sagra, Hist. de l'Ile de Cuba, Bot., p. 165. 1838-42. Bailey (1883); Krempelhuber (1880); Muell.-Arg. (1881, 1887c); Shirley (1889a). Q.

O. BONPLANDII var. OBLITERATA Muell.-Arg., Bull. Herb. Boiss. 1: 54. 1893. V.

O. CAVATA (Ach.) Muell.-Arg., Flora 65: 499. 1882. Thelotrema cavatum Ach., Kongl. Svenska Vetensk.-Akad. Nya. Handl. 33: 92. 1812. Muell.-Arg., (1882b). NSW.

*O. DIFFRACTELLA Muell.-Arg., Hedwigia 30: 50. 1891. Bailey (1891a); Muell.-Arg. (1891b); Shirley (1891b). Q.

*O. ENDOMELAENA Muell.-Arg., op. cit. 32: 131. 1893. Shirley (1893a). Q.

*O. GONIOSTOMA Muell.-Arg., op. cit. 30: 51. 1891. Bailey (1891a); Muell.-Arg. (1891b); Shirley (1891b). Q.

*O. GYROSTOMOIDES Muell.-Arg., Flora 71: 47. 1888. Muell.-Arg. (1888a, 1893); Shirley (1889a). Q, V.

*O. JUGALIS Muell.-Arg., Bull. Herb. Boiss. 3: 313. 1895. Muell.-Arg. (1895c); Shirley (1896). Q.

O. LEUCOTYLIA (Nyl.) Muell.-Arg., Hedwigia 30: 51. 1891. Thelotrema leucotylium Nyl., Bull. Soc. Linn. Normand. (2) 7: 166. 1873. Shirley (1889a, as "leucostyla," 1893a).

O. MICROPORELLA (Nyl.) A. Zahlbr., Cat. Lich. Univ. 2: 595. 1923. Thelotrema microporellum Nyl., Ann. Sci. Nat., Bot. (4) 19: 327. 1863. According to Muell.-Arg., reports of this pertain to T. australiense.

*O. OCTOLOCULARIS (Knight) Shirley, Proc. Roy. Soc. Queensland 6: 188. 1889. Ascidium octoloculare Knight ex Bailey, op. cit. 1: 152. 1884. Porina octolocularis Knight ex Shirley (pro synon.), op. cit. 6: 188. 1889. Q.

*O. PHLYCTIDIOIDES Muell.-Arg., Hedwigia 32: 130. 1893. Shirley (1893a). Q.

*O. PLATYCHLAMYS Muell.-Arg., Bull. Herb. Boiss. 3: 313. 1895. Shirley (1896). Q.

*O. PULCHRA Muell.-Arg., Nuovo Giorn. Bot. Ital. 23: 395. 1891. Shirley (1891b). Q.

*O. TEREBRATULA Muell.-Arg., Mém. Soc. Phys. Hist. Nat. Genève 29 (8): 12. 1887. Thelotrema terebratulum Nyl., Ann. Sci. Nat., Bot. (5) 7: 315. 1867. Bailey (1881, 1883); Muell.-Arg. (1881); Shirley (1889a). Q.

*. VIOLACEA Raes., Arch. Soc. Zool. Bot. Fenn. "Vanamo" 3: 184. 1949. Q.

*O. VIOLACEA var. GLAUCA Raes., l.c. p. 185. 1949. Q.

*O. VIRIDIPALLENS Muell.-Arg., Flora 70: 397. 1887. Shirley (1889a). Q.

*O. XANTHOLEUCA Muell.-Arg., Hedwigia 30: 51. 1891. Bailey (1891a); Shirley (1891b). Q.

*O. ZEORINA Muell.-Arg., Nuovo Giorn. Bot. Ital. 23: 394. 1891. Shirley (1891b). Q.

OCHROLECHIA

O. PALLESCENS (L.) Mass., Nuovi Ann. Sci. Nat. 7: 212. 1853. Lichen pallescens L., Sp. Pl. p. 1142. 1753. Lecanora parella var. pallescens Ach., Lichenogr. Univ. p. 370. 1810. Lecanora pallescens Roehl., Deutschl. Fl. 3(2): 76. 1813. Bailey (1881): Knight (1882); Muell.-Arg. (1892b, 1893); Shirley (1889a). NSW, Q, V, WA.

O. PALLESCENS var. TURNERI (Sm.) Koerb., Syst. Lich. German. p. 133. 1882. Lichen turneri Sm. & Sowerby, Engl. Bot. 12: t. 857. 1801. Wilson (1887). V.

O. PARELLA (L.) Mass., Ricerch. Auton. Lich. p. 32. 1852. Lichen parellus L., Mantissa 1: 132. 1767. Lecanora parella Ach., Lichenogr. Univ. p. 370. 1810. Bailey (1883); Hampe (1853); F. Mueller (1881); Shirley (1889a); Stirton (1899b); Wilson (1889c, 1890b). Q, V, WA.

O. PARELLA f. PHLOEOLEUCA (Nyl.) A. Zahlbr., Cat. Lich. Univ. 5: 690. 1928. Lecanora parella f. phloeoleuca Nyl. ex Cromb., J. Linn. Soc. London, Bot. 15: 440. 1876. Bailey (1881); Shirley (1889a). Q. Verseghy (1962) regards this as a Pertusaria species.

O. TARTAREA (L.) Mass., Ricerch. Auton. Lich. p. 30. 1852. Lichen tartareus L., Sp. Pl. p. 1141. 1753. Bailey (1881, 1883); F. Mueller (1881); Shirley (1889a). Q.

OPEGRAPHA

O. ATRA Pers., Ann. Bot. (Usteri) 1: 30. 1794. Wilson (1888). V.

O. ATRA var. DENIGRATA (Ach.) Schaer., Lich. Helvet. Spicil. 1: 48. 1823. O. denigrata Ach., Method Lich. p. 27. 1803. Wilson (1888, as "f. abbreviata"). V.

O. ATRA f. CERASI (Chev.) Arn., Flora 67: 662. 1884. O. cerasi Chev., J. Phys. Chem. Hist. Nat. 94: 38. 1822. O. atra var. parallela Light. ex Mudd, Man. Brit. Lich. p. 232. 1861. Wilson (1888). V.

O. BONPLANDII Fée, Essai Crypt. Ecorc. Officin. p. 25. 1824. Muell.-Arg. (1891a, 1893); Shirley (1889a). Q.

O. BONPLANDII var. ABBREVIATA (Fée) Muell.-Arg., Mém. Soc. Phys. Hist. Nat. Genève 29 (8): 17. 1887. O. abbreviata Fée, Essai Crypt. Ecorc. Officin. p. 25. 1824. Shirley (1893a). Q.

O. CONTEXTA Stirt., Grevillea 3: 35. 1874. Wilson (1888). V.

O. DIAPHORA (Ach.) Ach., Method. Lich., p. 19. 1803. Lichen diaphorus Ach., Lichenogr. Suec. Prodrom. p. 20. 1798. Opegrapha varia var. diaphora Fr., Lichenogr. Europ. Reform. p. 365. 1831. Muell.-Arg. (1893); Shirley (1893a). Q, V.

*O. GROSSULINA Muell.-Arg., Nuovo Giorn. Bot. Ital. 23: 396. 1891. Shirley (1891a). Q.

*O. INALBESCENS (Stirt.) Muell.-Arg., Nuovo Giorn. Bot. Ital. 23: 397. 1891. Lecidea inalbescens Stirt., Trans. Proc. Roy. Soc. Victoria 17; 72. 1881. Bailey (1881, 1883); F. Mueller (1881); Muell.-Arg. (1891a); Shirley (1889a). Q.

*O. INTERVENIENS Muell.-Arg., Hedwigia 30: 53. 1891. Shirley (1891b). Q.

*O. INTRUSA Stirt., Trans. Proc. Roy. Soc. Victoria 17: 73. 1881. Bailey (1881, 1883); F. Mueller (1881); Shirley (1889a). Q.

*O. LACTEELLA Muell.-Arg., Bull. Herb. Boiss. 1: 55. 1893. Shirley (1896). V.

*O. LEPTOCARPOIDES A. Zahlbr., Cat. Lich. Univ. 2: 213. 1923. Opegrapha leptocarpa Muell.-Arg., Bull. Herb. Boiss. 3: 317. 1895 (non Mont., 1838-1842). V.

*O. LEUCINIA Muell.-Arg. ex Shirley, Proc. Roy. Soc. Queensland 6: 195. 1889. Q.

*O. MEGAGONIDIA Knight, Trans. Linn. Soc. London, Bot. 2: 43. 1882. NSW.

*O. MINUTULA Muell.-Arg., Bull. Herb. Boiss. 3: 317. 1895. Shirley (1896). Q.

O. PERSOONII Ach., Method. Lich., p. 17. 1803. O. saxicola var. persoonii Stizenb., Nova Actorum Acad. Caes. Leop.-Carol. Nat. Cur. 32: 30. 1865. Wilson (1888). V.

O. PROSODEA Ach., Method. Lich. p. 22. 1803. Muell.-Arg. (1893); Shirley (1896). V.

O. PUIGGARII Muell.-Arg., Flora 63: 42. 1880. Santesson (1952). Q.

*O. PULICARIS (Willd.) Schrad. var.? glomerulans A. Zahlbr., Cat. Lich. Univ. 2: 238. 1923. Opegrapha varia var. glomerulans Muell.-Arg., Bull. Herb. Boiss. 1: 56. 1893. V.

O. SIDERELLA Ach., Method. Lich., p. 25. 1803. O. vulgata var. siderella Nyl., Act. Soc. Linn. Bordeaux 21: 405. 1856. Shirley (1893a, as O. vulgata var. "subsiderella" Nyl.). Q.

O. SORORIELLA Muell.-Arg., Proc. Roy. Soc. Edinburgh 11: 467. 1888. Muell.-Arg. (1893). V.

O. TURNERI Leight., Ann. Mag. Nat. Hist. (2) 13: 202. 1854. Hue (1890-92); Muell.-Arg. (1893); Nylander (1886). NSW, V.

O. VARIA Pers., Ann. Bot. (Usteri) 1: 30. 1794. Wilson (1888). V.

O. VARIA var. HETEROCARPA Muell.-Arg., Flora 71: 512. 1888. Muell.-Arg. (1893); Shirley (1894). V.

O. VULGATA (Ach.) Ach., Method. Lich. p. 20. 1803. Lichen vulgatus Ach., Lichenogr. Suec. Prodrom. p. 21. 1798. Muell.-Arg. (1893). V.

(Opegrapha)

*O. VULGATA var. PARALLELA Muell.-Arg., Bull. Herb. Boiss. 1: 56. 1893. V.

PANNARIA

*P. AENEA Muell.-Arg., Bull. Herb. Boiss. 4: 92. 1896. Muell.-Arg. (1898). Q.

*P. BRISBANENSIS Knight ex Shirley, Lich. Fl. Queensland, part 4: 194. 1890. Q.

P. CHEIROLEPIS F. Wils., Victoria Naturalist 6: 61. 1889, nomen nudum.

*P. ELATIOR Stirt. ex Bailey, Queensland Agr. J. 5: 486. 1899. Q.

*P. FLEXUOSA Knight ex Shirley, Syn. Queensland Fl. II Suppl. p. 82. 1888. Shirley (1889a). Q.

P. FULVESCENS (Mont.) Nyl., Mém. Soc. Scienc. Nat. Cherbourg 5: 109. 1857. Parmelia fulvescens Mont., Ann. Sci. Nat., Bot. (3) 10: 1848. Bailey (1891a); Krempelhuber (1880); F. Mueller (1881); Muell.-Arg. (1891b); Wilson (1887, 1889c). NSW, Q, V.

P. LURIDA (Mont.) Nyl., Mém. Soc. Scienc. Nat. Cherbourg 5: 109, 1857. Collema luridum Mont., Ann. Sci. Nat., Bot. (2) 18: 236. 1842. Pannaria sublurida Nyl., op. cit. (4) 11: 256. 1859. Wilson (1888, 1889c). Q, V.

P. MARIANA (Fries) Muell.-Arg., Flora 60: 321. 1887. Parmelia mariana Fr., Syst. Orb. Veget. 1: 284. 1825. Pannaria pannosa Nyl., Mem. Soc. Scienc. Nat. Cherbourg 3: 176. 1855. Bailey (1891a); Krempelhuber (1880); F. Mueller (1881); Muell.-Arg. (1891b); Shirley (1889a); Turner (1905). NSW, Q.

*P. MARIANA var. ACCOLENS (Stirt.) A. Zahlbr., Cat. Lich. Univ. 3: 247. 1925. Pannaria pannosa var. accolens Stirt. ex Bailey, Queensland Agr. J. 5: 40. 1889. Q.

P. MARIANA var. FUNEBRIS (Kremp.) Vain., Ann. Acad. Sci. Fenn. ser. A, 15 (6): 8. 1921. Pannaria funebris Kremp., J. Mus. Godeffroy 1 (4): 101. 1873. Q.

P. MARIANA f. ISIDIOIDEA Muell.-Arg., Flora 70: 321. 1887. Shirley (1889a, 1893a). Q.

P. MOLYBDODES F. Wils., Victoria Naturalist 6: 61. 1889, nomen nudum.

*P. MYRIOLOBA Muell.-Arg., Bull. Herb. Boiss. 4: 92. 1896. NSW. Bibby (1954); Muell.-Arg. (1898). NSW, Q.

"P. NIGRA Hud.," Wilson, Victoria Naturalist 4: 87. 1887, nomen nudum. Probably refers to Placynthium nigrum (Lichen niger Huds.).

*P. NIGRATA Muell.-Arg., Bull. Herb. Boiss. 4: 91. 1896. Muell.-Arg. (1898). V.

*P. OBSCURA Muell.-Arg., Bull. Herb. Boiss. 4: 91. 1896. V.

P. PARMELIAE F. Wils., Victoria Naturalist 6: 61. 1889, nomen nudum.

P. RUBIGINOSA (Thunb. ex Ach.) Del., Dict. Class. Hist. Nat. 13: 20. 1828. Lichen rubiginosus Thunb. ex. Ach., Lichenogr. Suec. Prodrom. p. 99. 1798. Coccocarpia rubiginosa Hampe, Linnaea 28: 217. 1856. Crombie (1880); Fries (1846-47); Hue (1890-92); Krempelhuber (1880); F. Mueller (1881); Shirley (1889e). NSW, Q, V, WA.

P. RUBIGINOSA var. LANUGINOSA (Hoffm.) A. Zahlbr., Cat. Lich. Univ. 3: 258. 1925. Lichen lanuginosus Hoffm., Enum. Lich. p. 82. 1784. Shirley (1889a). Q.

*P. SOREDIATA Knight ex Bailey, Proc. Roy. Soc. Queensland 1: 90. 1884. Shirley (1889a). Q.

*P. SUBIMMIXTA Knight var. RECEDENS Muell.-Arg., Bull. Herb. Boiss. 4: 92. 1896. Muell.-Arg. (1898). V.

*P. TERRESTRIS Stirt. ex Bailey, Queensland Agr. J. 5: 486. 1899. Q.

P. THRAUSTOLEPIS F. Wils., Victoria Naturalist 6: 61. 1889, nomen nudum.

PARATHELIUM

*P. DECUMBENS Muell.-Arg., Hedwigia 32: 134. 1893. Muell.-Arg. (1895b); Shirley (1893a). Q.

PARMELIA

P. ADPRESSA Kremp., Flora 59: 72. 1876. Muell.-Arg. (1891a, 1892b); Shirley (1893a). Q. WA.

P. ALPICOLA Th. Fr., Nova Acta Reg. Soc. Scient. Upsal. (3) 3: 157. 1861. Wilson (1890a). V.

*P. AMPHIXANTHA Muell.-Arg., Flora 71: 139. 1888. V. Bibby & Smith (1954); Muell.-Arg. (1892b); Vainio (1900); Filson (1967). V., WA.

P. ASPERA Mass., Mem. Lichenogr. p. 53. 1853. P. olivacea var. exasperata Nyl. Mém. Soc. Imp. Sci. Nat. Cherbourg 5: 105. 1857. Wilson (1887). V.

*P. AUSTRALIENSIS Cromb., J. Linn. Soc. London, Bot. 17: 395. 1879. Bibby (1950); Hue (1890-92); Kurokawa (1969); Willis (1959b); Wilson(1887, "var. isidiophora," nomen nudum). SA, V, WA.

P. AUSTROSINENSIS A. Zahlbr., Symbol. Sinicae 3: 192. 1930. Hale (1965). Q.

P. BORRERI Turner in Sm. & Sowerby. Engl. Bot. 25: tab. 1780. 1807. Wilson (1887). V.

*P. BORRERI var. CORALLOIDEA Muell.-Arg., Flora 70: 60. 1887. P. isabellina Kremp., Verh. zool.-bot. Ges. Wien 30: 338. 1880 (pr. p.). F. Mueller (1881); Muell.-Arg. (1887c). NSW, V.

P. BORRERI var. ULOPHYLLA (Ach.) Nyl., Flora 55: 547. 1872. P. caperata var. ulophylla Ach., Lichenogr. Univ. pl 458. 1810. P. ulophylla F. Wils., Pap. Proc. Roy. Soc. Tasmania for 1892, p. 172, 1893. Wilson (1889a,b,c, 1890b). NSW, V, WA.

*P. BRISBANENSIS Stirt., Trans. Proc. Roy. Soc. Victoria 17: 69. 1881. Bailey (1881, 1883); F. Mueller (1881); Muell.-Arg. (1891a); Shirley (1889a); Stirton (1899a). Q.

P. CAMTSCHADALIS Eschw. in Martius, Fl. Brasil 1: 202. 1833. Wilson (1889a). V.

P. CAPERATA (Hoffm.) Ach., Method. Lich. p. 216. 1803. Platisma caperatum Hoffm., Descr. Adumbr. Plant. Lich. 2: 50. 1794. Bailey (1883); Fries (1846-47); Hampe (1853); Hellbom (1896); Krempelhuber (1880);

F. Mueller (1881); Nylander (1857); Shirley (1889a, 1892b); Stirton (1899a); Watts (1903); Weber (1969); Wilson (1889a, var. somediosa, nomen, 1889b,c, 1890b); Zahlbruckner (1896). V.

P. CAPERATULA (Nyl.) Nyl., Flora 68: 606. 1885. P. caperata var. caperatula Nyl., Syn. Lich. 1: 377. 1860. Bailey (1881); Hue (1890-92); Muell.-Arg. (1881); Shirley (1889a); Stirton (1899b); Wilson (1887). Q, V.

P. CETRARIOIDES Del. in Duby, Bot. Gallic. 2: 601. 1830. Stirton (1899b). Q.

P. CETRATA Ach., Syn. Lich. p. 198. 1814. Maiden (1898); Watts (1903). NSW.

*P. CHEELII Gyeln., Ann. Mycol. 36: 271. 1938. NSW.

*P. CONFERTULA Stirt., Trans. Proc. New Zealand Inst. 32: 77. 1899. Q.

P. CONGRUENS Ach., Lichenogr. Univ. p. 491. 1810. (=P. molliuscula Ach. fide Zahlbruckner, Cat. Lich. Univ. 6: 72. 1929). Muell.-Arg. (1892b). WA.

*P. CONIOCARPA Laur., Linnaea 2: 39. 1827. Sine loc.

P. CONSPERSA (Ach.) Ach., Method. Lich. p. 205. 1803. Lichen conspersus Ach., Lich. Suec. Prodrom. p. 118. 1798. Bailey (1881, 1883); Bibby & Smith (1954); Crombie (1880); Fries (1846-47); Hampe (1853); Knight (1882); Maiden (1898); F. Mueller (1858); Muell.-Arg. (1881, 1887c, 1892b); Shirley (1889a,d); Smith (1962); Turner (1905); Watts (1903); Willis (1953). NSW, Q, SA, WA.

P. CONSPERSA var. AUSTROAFRICANA (Stirt.) Stizenb., Ber. Thaetigk. S. Gallischen Naturwiss. Ges. 1888-1889, p. 152. 1890. P. austroafricana Stirt., Trans. Glasgow Soc. Field Nat. 5: 212. 1877. Bailey (1881); F. Mueller (1881); Stirton (1899a). Q, V.

*P. CONSPERSA var. CAESPITOSA Muell.-Arg., Bull. Herb. Boiss. 4: 90. 1896. Muell. Arg. (1898). V.

P. CONSPERSA var. CONSTRICTANS (Nyl. ex Cromb.) Muell.-Arg., Flora 66: 48. 1883. P. constrictans Nyl. ex Cromb. J. Linn. Soc. London, Bot. 15: 168. 1875. V.

P. CONSPERSA var. ERADICATA (Nyl.) Muell.-Arg., Flora 66: 48. 1883. Parmelia constrictans var. eradicata Nyl, ex Cromb. J. Linn. Soc. London, Bot. 15: 168. 1875. P. subconspersa var. eradicata A. Zahlbr., Annal. K. K. Naturh. Hofmus. 11: 195. 1895. NSW, V.

*P. CONSPERSA f. EXASPERATA Muell.-Arg., Flora 66: 47. 1883. Shirley (1890). Q.

P. CONSPERSA var. HYPOCLISTA Nyl., Syn. Lich. 1: 391. 1860. Muell.-Arg. (1892a,b, 1893); Zahlbruckner (1896). Q, V, WA.

*P. CONSPERSA var. HYPOCLISTOIDES Muell.-Arg., Flora 66: 201. 1883. V. *P. scabrosa Tayl. ex Hook., London J. Bot. 6: 162. 1847. WA. Muell.-Arg. (1888c, 1891a); Shirley (1891a). Q, V, WA.

*P. CONSPERSA var. INCISA (Tayl. ex Hook.) A. Zahlbr., Cat. Lich. Univ. 6: 132. 1929. *Parmelia incisa Tayl. ex Hook., London J. Bot. 6: 162. 1847. WA. P. conspersa f. laxa Muell.-Arg., Flora 66: 47. 1883. Muell.-Arg. (1888c, 1892a); Wilson (1890a). V, WA.

P. CONSPERSA f. ISIDIATA Anzi, Cat. Lich. Sondr. p. 28. 1860. P. conspersa var. corallina Kremp., Denkschr. Koenigl. Bayr. Bot. Ges. Regensburg 4(2): 135. 1861. P. conspersa f. isidiigera Hellb., Bih. Kongl. Svenska Vetensk.-Akad. Handl. 21 (III, No. 21): 47. 1896. Muell.-Arg. (1883a, 1892b). WA.

*P. CONSPERSA f. ISIDIOPHORA Muell.-Arg., Flora 66: 48. 1883. V.

P. CONSPERSA var. ISIDIOSA Nyl. Flora 64: 450. 1881. Krempelhuber (1880); Muell.-Arg. (1892b); Wilson (1887); Zahlbruckner (1896). NSW, V, WA.

*P. CONSPERSA var. MULTIPARTITA R. Br. ex Cromb., J. Linn. Soc. London, Bot. 17: 394 (1879) 1880. NSW.

*P. CONSPERSA var. NIGROMARGINATA Stirt., Trans. Proc. New Zealand Inst. 32: 78. 1899. V.

*P. CONSPERSA var. POLYPHYLLOIDES Muell.-Arg., Flora 66: 47. 1883. SA.

P. CONSPERSA var. STENOPHYLLA Ach., Meth. Lich. p. 206. 1803. Crombie (1880); Muell.-Arg. (1892b); Wilson (1887); Zahlbruckner (1896). V, WA.

*P. CONSPERSA var. STENOPHYLLOIDES Muell.-Arg., Hedwigia 31: 193. 1892. WA. Muell.-Arg. (1893). V, WA.

P. CONSPERSA var. SUBCONSPERSA (Nyl.) Stein in Meyer, Ostafr. Gletscherfahrten, p. 316. 1890. P. subconspersa Nyl., Flora 52: 293. 1869. Krempelhuber (1880); Muell.-Arg. (1887c); F. Mueller (1881). V.

*P. CONVOLUTA Kremp., Verh. zool.-bot. Ges. Wien 30: 337. 1880. Muell.-Arg. (1881); Tate (1881). CA, SA.

*P. CORRUGIS (Fries) Muell.-Arg. f. SOREDIATA Muell.-Arg., Flora 71: 202. 1888. Muell.-Arg. (1891a); Shirley (1891a). Q.

P. CRINITA Ach., Syn. Lich. p. 196. 1814. Hale (1965). Q.

*P. DICHOTOMA Muell.-Arg., Flora 69: 257. 1886. NSW. Kurokawa (1969); Weber (1971). ACT, V.

P. DILATATA Vain., Acta Soc. Faun. Flora Fenn. 7(7): 33. 1890. Hale (1965). NSW, Q.

P. ECILIATA Nyl. in Fourn., Mexic. Plant. 1: 3. 1872. P. crinita var. eciliata Nyl., Flora 52: 291. 1869. P. latissima var. eciliata Nyl. in Fourn., Mexic. Plant. 1: 3. 1872. Muell.-Arg. (1891a, 1892a, orthogr. error, "ciliata"); Shirley (1891a, 1892a). Q.

*P. ERUBESCENS Stirton, Scott. Naturalist 4: 201. 1877-1878. Bailey (1881, 1883); F. Mueller (1881); Shirley (1889a); Stirton (1899a). Q.

*P. ERUMPENS Kurok., Lich. Rar. Crit. Exsicc. 74. 1969. *P. tenuirima Hook. & Tayl. f. corallina Muell.-Arg., Flora 66: 46. 1863. Q. Muell.-Arg. (1883a, 1892a); Shirley (1889b, 1892b); Wilson (1887); Zahlbruckner (1896, 1904b). NSW, Q, V.

*P. EUPLECTA Stirt., Scott. Naturalist 4: 299. 1877-78. Bailey (1881, 1883); F. Mueller (1881); Shirley (1889a); Stirton (1889a). Q.

*P. EXORIENS Stirt., Trans. Proc. New Zealand Inst. 32: 75. 1899. Q.

*P. FERAX Muell.-Arg., Flora 69: 257. 1886. NSW. Vainio (1900); Kurokawa (1967).

(Parmelia)

*P. FILARSZKYANA Gyeln., Ann. Mycol. 36: 278. 1938. NSW.

*P. FLAVESCENTIREAGENS Gyel., Fedde Repert. 36: 154. 1934. NSW. Kurokawa (1969).

P. FULIGINOSA Nyl., Flora 51: 346. 1868. P. dendritica var. fuliginosa Muell.-Arg., Hedwigia 22: 193. 1892. WA.

*P. FURCATA Muell.-Arg., Flora 69: 256. 1886. NSW. *P. foliosa Kremp. ex Gyeln., Ann. Mus. Nat. Hung. (Bot.) 29: 35. 1935. NSW. Kurokawa (1969b).

P. GALBINA Ach., Synops. Method. Lich. p. 195. 1814. *P. endoleuca Tayl. ex Hook., London J. Bot. 6: 167. 1847. WA. Muell.-Arg. (1888c); Weber (1969). ACT, WA.

*P. GRACILIS Muell.-Arg., Flora 70: 317. 1887. P. gracilenta Vain., Mem. Herb. Boiss. 5: 6. 1900. Shirley (1889b). Q.

P. GRAYANA Hue, Nouv. Arch. Mus. Paris (4) 1: 184. 1899. Kurokawa (1969a). NSW.

*P. HETEROCHROA Hale & Kurok., Contr. U.S. Natl. Herb. 36: 154. 1964. P. hypoxantha Stirt., Trans. Proc. New Zealand Inst. 32: 76. 1899 (non Muell.-Arg., 1881). P. tiliacea var. stenophylla Muell.-Arg., Flora 66: 46. 1883. P. tiliacea var. rugulata Muell.-Arg., Nuovo Giorn. Bot. Ital. 23: 388. 1891. *P. tiliacea var. convexula Muell.-Arg., Bull. Herb. Boiss. 4: 90. 1896. Q. Muell.-Arg. (1898); Shirley (1889b, 1891b, 1893a); Stirton (1899b). Q.

*P. HOSPITANS Muell.-Arg., Flora 66: 76. 1883. NSW, V.

*P. HYPOPROTOCETRARICA Kurok. & Elix, J. Jap. Bot. 46: 113. 1971. ACT.

*P. INSINUATA Nyl., Flora 69: 324. 1886. P. sphaerospora Knight, Trans. Linn. Soc. London, Bot. 2: 49. 1882 (non P. sphaerospora Nyl.). Hue (1890-92); Shirley (1889a). NSW.

*P. JELINEKII Kremp., Reise Oesterr. Fregatta "Novara," Bot. 1: 114. 1870. Sine loc.

P. LABROSA (A. Zahlbr.) M. Hale, J. Jap. Bot. 43: 325. 1968. P. tenuirima var. labrosa A. Zahlbr., Denkschr. Kaiserl. Akad. Wiss. Math.-Naturwiss. Kl. 104: 108. 1941. N.S.W. Coolamon Plain, McVean 6490 (COLO).

*P. LACERATULA Nyl., Mem. Soc. Imp. Sci. Nat. Cherbourg 5: 105. 1857. Hale (1960). Q.

P. LAEVIGATA (Sm. in Sm. & Sowerby) Ach. Syn. Lich. p. 212. 1814. Lichen laevigatus Sm. in Sm. & Sowerby, Engl. Bot. 26: tab. 1852. 1808. F. Mueller (1881). Q.

P. LATISSIMA Fée, Suppl. Essai Cryptog. Ecorc. Officin. p. 119. 1837. Bailey (1883); Krempelhuber (1868, 1880); F. Mueller (1881); Muell.-Arg. (1887c); Shirley (1889a). NSW, Q.

P. LATISSIMA var. CORNICULATA Kremp., Flora 61: 463. 1878. F. Mueller (1887).

P. LATISSIMA var. SOREDIATA Nyl. ex Hue, Nouv. Arch. Mus. Paris (3) 2: 282. 1890. Bibby (1954). Q.

*P. LIMBATA Laur., Linnaea 2: 39. 1827. Bailey (1881, 1883); Baker (1902); Crombie (1880); Hue (1890-92); Krempelhuber (1868, 1880); Kurakawa (1965); F. Mueller (1881); Muell.-Arg. (1887c, 1891a); Shirley (1889a, 1892b, 1893a); Stirton (1899b); Turner (1905); Zahlbruckner (1896, 1910). NSW, Q, V.

*P. LIMBATA f. ENDOCOCCINEA Muell.-Arg., Flora 70: 318. 1887. Shirley (1889b). Q.

*P. MEIZOSPORA (Nyl.) Nyl., Flora 52: 292. 1869. P. tiliacea var. meizospora Nyl., Syn. Lich. 1: 382. 1860. Knight (1882); Shirley (1889a, 1893a). NSW, Q.

P. MEIZOSPORA Nyl. f. ISIDIOSA Muell.-Arg., Flora 67: 620. 1884. (=P. amazonica Nyl., fide Zahlbr., Cat. Lich. Univ. 6: 151. 1929). Muell.-Arg. (1892a). Q.

*P. METAMORPHOSA Gyel., Ann. Mycol. 36: 284. 1938. NSW. Kurokawa (1969). NSW.

P. MOUGEOTII Schaer. ex Dietr., Deutschl. Kryptog. Gewaechse, 4 Abt. p. 118. 1846. Knight (1882); Shirley (1889a); Wilson (1890a). NSW, Q, V.

*P. NEOHOLLANDICA Nyl., Flora 69: 324. 1886. Hue (1890-92). Sine loc.

P. NILGHERRENSIS Nyl., op. cit. 57: 318. 1874. Hue (1890-92); Nylander (1885). Sine loc.

*P. NITESCENS Stirt., Scott. Naturalist 4: 299. 1877-78. Bailey (1881, 1883); F. Mueller (1881); Shirley (1889a). Q.

*P. NOTATA Kurok., J. Jap. Bot. 46: 33. 1971. NSW.

*P. OBVERSA Stirt., Trans. Proc. New Zealand Inst. 32: 76. 1899. Sine loc.

P. OLIVACEA (L.) Ach. Meth. Lich. p. 213. 1803. Lichen olivaceus L., Sp. Pl. p. 1143. 1753. Nylander (1857); Shirley (1889a); Wilson (1887, 1890b). Q, V, WA. Most likely a misidentification of this boreal species.

P. OLIVARIA (Ach.) Th. Fr., Lichenogr. Scand. 1: 112. 1871. P. perlata var. olivaria Ach., Meth. Lich. p. 217. 1803. Bailey (1881, 1883); F. Mueller (1881); Muell.-Arg. (1891a); Shirley (1888b, 1889a, 1891a); Wilson (1887). Q, V.

P. PERFORATA (Wulf. in Jacq.) Ach., Method. Lich. p. 217. 1803. Lichen perforatus (Wulf. in Jacq., Collect. Bot. 1: 116. 1786. Bailey (1881); Baker (1902); Fries (1846-47); Hampe (1853); Krempelhuber (1868, 1880); F. Mueller (1881); Muell.-Arg. (1887c); Nylander (1857); Shirley (1889a); Stirton (1899b); Turner (1905); Willis (1953); Zahlbruckner (1896). Q, NSW, SA.

P. PERFORATA var. CILIATA Nyl., Syn. Lich. 1: 377. 1860. Muell.-Arg. (1891a); Shirley (1891a); Wilson (1887). Q, V.

P. PERFORATA var. ULOPHYLLA Mey. & Flot., Nova Actorum Acad. Caes. Leop.-Carol. Nat. Cur. 19, Suppl. p. 218. 1843. *P. concors Kremp., Verh. zool.-bot. Ges. Wien 30: 337. 1880. Muell.-Arg. (1887c, 1892a). V.

*P. PERLATA (Huds.) Ach., Method. Lich. p. 216. 1803. Lichen perlatus Huds., Fl. Angl. p. 448. 1762. Bailey (1881, 1883); Bibby & Smith (1954); Fries (1846-47); Hale (1961, 1965); Hampe (1853); Hue

(1890-92); Krempelhuber (1880); F. Mueller (1881); Shirley (1889a); Smith (1962); Weber (1969); Wilson (1889c). NSW, Q, V, WA.

*P. PERLATA f. ASPERA Muell.-Arg., Ann. K.K. Naturhist. Hofmus. 7: 303. 1892. P. proboscidea var. aspera Muell.-Arg., Flora 69: 256. 1886. NSW.

P. PERLATA var. CILIATA (DC. in Lam. & DC.) Duby, Bot. Gall. 2: 601. 1830. Lobaria perlata var. ciliata DC. in Lam. & DC., Fl. Franç. ed. 3. 2: 403. 1805. Muell.-Arg. (1892a); Nylander (1857); Shirley (1893a); Wilson (1887). Q, V.

*P. PERLATA f. CORALLINA (Muell.-Arg.) Muell.-Arg., Ann. K.K. Naturhist. Hofmus. 7: 303. 1892. P. proboscidea var. corallina Muell.-Arg., Flora 67: 616. 1884. NSW.

P. PERLATA f. SOREDIIFERA Muell.-Arg., op. cit. 74: 382. 1891. Knight (1882); Hue (1890-92). NSW.

*P. PERMUTATA Stirt., Scott. Naturalist 4: 252. 1877-78. Bailey (1881, 1883); F. Mueller (1881); Shirley (1889a); Stirton (1899a). Q.

P. PERRUGATA Nyl., Flora 68: 295. 1885. P. prolixa f. perrugata Harm., Bull. Soc. Scienc. Nancy (2) 31: 229. 1897.

*P. PLATYCARPA Stirt., Scott. Naturalist 4: 252. 1877-78. Bailey (1881-1883); Hale (1965, treated as a nomen nudum, the type material impossible to interpret); F. Mueller (1881); Shirley (1889a); Stirton (1899a). Q.

P. PROBOSCIDEA Tayl. in Mack., Fl. Hibern. 2: 143. 1836. Muell.-Arg. (1884b). Q.

*P. PRUINATA Muell.-Arg., Flora 66: 46. 1883. Kurokawa (1969b); Shirley (1890); Vainio (1900). Q.

*P. PUBESCENS F. Wils., Victoria Naturalist 6: 180. 1890. WA. Nomen nudum.

P. PULLA (Schreb.) Ach., Syn. Lich. p. 206. 1814. Lichen pullus Schreb., Spicil. Fl. Lips. p. 131. 1771. P. olivacea var. prolixa Ach., Method. Lich. p. 214. 1803. P. prolixa Roehl., Deutsch. Fl. 3: 100. 1813. P. dendritica Pers., Ann. Wetterau. Ges. 2: 16. 1811. *P. imitatrix Tayl. ex Hook., London J. Bot. 6: 161. 1847. WA. *P. subprolixa var. angusta Kremp., Verh. zool.-bot. Ges. Wien 30: 337. 1880. V. Bibby & Smith (1954); Hellbom (1896); Hue (1890-92); F. Mueller (1881); Muell.-Arg. (1883a, 1892b); Tate (1881); Weber (1969); Wilson (1887). NSW, SA, V, WA.

(P. PULLA) P. PROLIXA var. SUBPROLIXA (Kremp.) A. Zahlbr., Cat. Lich. Univ. 6: 107. 1929. P. subprolixa Kremp., Verh. zool.-bot. Ges. Wien 30: 337. 1880. P. imitatrix var. subprolixa Muell.-Arg., Flora 66: 47. 1883. F. Mueller (1881); Muell.-Arg. (1887c). NSW, SA, V.

P. QUERCINA (Willd.) Vain., Termesz. Fuezetek 22: 279. 1899. Lichen quercinus Willd., Fl. Berol. Prodr. p. 353. 1787. P. tiliacea Ach., Method. Lich. p. 215. 1803. Crombie (1880); Nylander (1857); Shirley (1889a); Stirton (1899b); Wilson (1887, 1889c, 1890b). NSW, Q, V, WA.

*P. QUERCINA var. AFFIXA (Stirt. ex Bailey) A. Zahlbr., Cat. Lich. Univ. 6: 191. 1929. P. tiliacea var. affixa Stirt. ex Bailey, Queensland Agric. J. 5: 485. 1899. Q.

(Parmelia)

P. QUERCINA f. MINOR (Kremp.) A. Zahlbr., Cat. Lich. Univ. 6: 190. 1929. P. tiliacea f. minor Kremp., Reise Oesterr. Fregatte "Novara," Bot. 1: 115. 1870. Muell.-Arg. (1892a,b). WA.

*P. QUERCINA var. REGULATA (Muell.-Arg.) A. Zahlbr., Cat. Lich. Univ. 6: 191. 1929. P. tiliacea var regulata Muell.-Arg., Nuovo Giorn. Bot. Ital. 23: 388. 1891. Bailey (1891); Muell.-Arg. (1892a). Q.

P. QUERCINA var. RIMULOSA (Muell.-Arg.) A. Zahlbr., Cat. Lich. Univ. 6: 191. 1929. P. tiliacea var. rimulosa Muell.-Arg., Proc. Roy. Soc. Edinburgh 11: 458. 1882. Q.

P. QUERCINA var. SULPHUROSA (Tuck.) A. Zahlbr., Cat. Lich. Univ. 6: 192. 1929. P. tiliacea var. sulphurosa Tuck., Syn. N. Am. Lich. 1: 57. 1882. Muell.-Arg. (1883a); Shirley (1893a). Q.

P. RAMPODDENSIS Nyl., Acta Soc. Sci. Fenn. 26 (10): 7. 1900. Hale (1965); Kurokawa (1969b). NSW.

*P. REDACTA Stirt., Trans. Proc. New Zealand Inst. 32: 76. 1899. NSW.

P. RELICINA Fries, Syst. Orb. Veget. 1: 283. 1825. Hue (1890-92); Nylander (1857). Sine loc.

P. RECIPIENDA Nyl., Flora 68: 609. 1885. Kurokawa (1969). NSW. Hale (1965) synonymized this with P. subcaperata Kremp.

*P. REPARATA Stirt., Scott. Naturalist 4: 201. 1877-78. Q. *P. virens Muell.-Arg., Flora 69: 255. 1886. Q. Bailey (1881, 1883); Hale (1965); F. Mueller (1881); Shirley (1889a,b). Q.

P. RETICULATA Tayl. in Mack., Flora Hibern. 2: 148. 1836. P. cetrata var. sorediifera Vain., Etud. Lich. Brésil 1: 40. 1890. Muell.-Arg. (1892a); Shirley (1893a). Q, V.

P. REVOLUTA Flk. in Spreng., Syst. Veg. 1: 284. 1827. Bailey (1881, 1883); Hue (1890-92); Krempelhuber (1868); F. Mueller (1881); Shirley (1889a,d, 1892b); Stirton (1899b); Tate (1887). Q, SA.

P. ROYI Stirt., credited to Australia by Zahlbruckner, Cat. Lich. Univ. 6: 277, was described from Canada, not Australia, and is synonymous with Anzia colpodes.

*P. RUBRIREAGENS Gyeln., Ann. Mycol. 36: 288. 1938. Sine loc.

P. RUDECTA Ach., Syn. Lich. p. 197. 1814. *P. subrudecta var. australica Raes., Ann. Bot. Soc. Zool.-Bot. "Vanamo" 20(3): 3. 1944. NSW. Hale (1965a).

P. RUTIDOTA Hook. f. & Tayl., London J. Bot. 3: 645. 1844. *P. ochroleuca Muell.-Arg., Flora 65: 306. 1882 (non Ach.). NSW. Bibby & Smith (1954); Hale (1960); Hampe (1853); Muell.-Arg. (1883a, 1892b); Shirley (1889a); Smith (1962); Weber (1969); Willis (1953). NSW, Q, WA.

P. RUTIDOTA f. SOREDIOSA Muell.-Arg., Bull. Soc. Bot. Belg. 31: 30. 1892. Shirley (1893a). Q.

P. SAXATILIS (L.) Ach., Method. Lich. p. 204. 1803. Lichen saxatilis L., Sp. Pl. p. 1142. 1753. Darbishire (1912); Muell.-Arg. (1892a); Shirley (1889a); Zahlbruckner (1896). NSW, Q, V.

P. SAXATILIS var. AIZONII Del. in Duby, Bot. Gall. 2: 602. 1830. P. saxatilis var. furfuracea Linds., Trans. Roy. Soc. Edinburgh 12: 227. 1859. Wilson (1889a). V.

*P. SCABROSA Tayl., London J. Bot. 6: 162. 1847. WA. *P. amplexula Stirt., Trans. Proc. Roy. Soc. Victoria 17: 69. 1881. *P. conspersa var. polyphylloides f. exasperata Muell.-Arg., Flora 66: 47. 1883. *P. violascens Stirt., Trans. Proc. New Zealand Inst. 32: 77. 1899. *P. subexasperata Gyel., Fedde Repert. 29: 288. 1931. *P. lesdainiana Gyel., op. cit. 36: 161. 1934. *P. protoisidiata Gyel., Fedde Repert. loc. cit. p. 162. *P. subreagens Gyel., Ann. Mycol. 36: 291. 1938. Bailey (1881) Kurokawa (1969); F. Mueller (1881); Shirley (1889a); Stirton (1889a). NSW, Q, V, WA.

P. SCHWEINFURTHII Muell.-Arg., Proc. Roy. Soc. Edinburgh 11: 459. 1882. Muell.-Arg. (1887a). V.

P. SCHWEINFURTHII f. SOREDIATA Muell.-Arg., Flora 70: 59. 1882. P. protosorediata Gyeln., Fedde Repert. 29: 288. 1931. Muell.-Arg. (1892a). = P. perlata according to Hale (1965). V.

P. SCORTEA (Ach.) Ach., Method. Lich. p. 215. 1803. Lichen scorteus Ach., Lichenogr. Suec. Prodr. p. 119. 1798. P. tiliacea var. scortea Duby, Bot. Gall. 2: 601. 1830. Wilson (1887). V.

P. SCORTEA var. FERACISSIMA (Muell.-Arg.) Harm., Lich. de France 4: 559. 1910. P. tiliacea var. feracissima Muell.-Arg., Flora 69: 256. 1886. Muell.-Arg. (1886). NSW.

P. SIGNIFERA Nyl., Lich. Nov. Zel. p. 25. 1888. P. saxatilis var. signifera Muell.-Arg., Bull. Soc. Bot. Belg. 31: 30. 1892. Shirley (1894); Weber (1969). V.

P. SINUOSA (Sm.) Ach., Syn. Lich. p. 207. 1814. Lichen sinuosus Sm. in Sm. & Sowerby, Engl. Bot. 29.: tab. 2050. 1809. Bailey (1881; 1883); F. Mueller (1881); Shirley (1889a). Q.

*P. STRAMINEONITENS A. Zahlbr., Ann K. K. Naturhist. Hofmus. 11: 195. 1893. Zahlbruckner (1896). NSW.

P. SUBALBICANS Stirt., Scott. Naturalist 4: 254. 1877-78. *P. polycarpina A. Zahlbr., Cat. Lich. Univ. 6: 185. 1929. *P. polycarpa Tayl. ex Hook., London J. Bot. 6: 173. 1847 (non Spreng.) WA. Hue (1890-92); Maiden (1898); F. Mueller (1881); Muell-Arg. (1888c); Weber (1969). WA. Synonymy according to Hale in litt.

*P. SUBAMPHIXANTHA Gyeln., Ann. Mycol. 36: 290. 1938. V.

*P. SUBBRUNNEA Stirt., Trans. Proc. New Zealand Inst. 32: 80. 1899. V.

*P. SUBCOERULEA F. Wils., Victoria Naturalist 6: 69. 1889. V. Overlooked by Zahlbruckner, Cat. Lich. Univ.

*P. SUBDISTORTA Kurokawa, J. Hattori Bot. Lab. 32: 212. 1969. V.

P. SUBFLAVA Tayl. ex Hook, London J. Bot. 6: 174. 1847. P. laceratula Nyl., Syn. Lich. 1: 390. 1860. Hale (1960, 1965a); Hue (1890-92); F. Mueller (1881); Muell.-Arg. (1892a); Nylander (1857, nomen): Shirley (1889a); Stirton (1899b). Q.

*P. SUBFLAVA var. MINOR (Shirley) A. Zahlbr., Cat. Lich. Univ. 6: 214. 1929. P. laceratula var. minor Shirley, Proc. Roy. Soc. Queensland 8: 133. 1892. Shirley (1892b). Q.

P. SUBLAEVIGATA Nyl., Ann. Sci. Nat., Bot. (5) 7: 306. 1867. Parmelia hookeri Tayl. ex Hook. London J. Bot. 6: 169. 1847 (non Spreng. nec Fries). P. tiliacea var. sublaevigata Nyl., Syn. Lich. 1: 383. 1860. Shirley (1893a). Q.

*P. SUBNUDA Kurok., J. Jap. Bot. 46: 144. 1971. ACT.

P. SUBRUDECTA Nyl., Flora 69: 320. 1886. Hale (1965a); Weber (1971). NSW.

P. SUBRUGATA Kremp., Verh. zool.-bot. Ges. Wien 18: 320. 1868. *P. cyathina Stirt., Scott. Naturalist 4: 252. 1877-78. Bailey (1881, 1883); Hale (1965); F. Mueller (1881); Stirton (1899a); Shirley (1889a). Q.

*P. SUBSTRIGOSA Hale ex W. A. Web., Lich. Exsicc. COLO. no. 388. 1971. *P. conspersa var. strigosa Muell.-Arg., Bull. Herb. Boiss. 4: 90. 1896 (non P. strigosa Pers., 1826). V.

P. SUBTILIACEA Nyl., Flora 68: 614. 1885. Nylander (1886). NSW.

P. SUBTINCTORIA A. Zahlbr., Symb. Sinici 3: 193. 1930. P. virens f. isidiosa Muell.-Arg., Ann. K.K. Naturh. Hofmus. 7: 303. 1892. Weber (1969). NSW, Q.

P. SULPHURATA Nees & Flot., Linnaea 9: 501. 1835. Shirley (1889a). Q.

*P. SYDNEYENSIS Gyeln., Ann. Mycol. 36: 292. 1938. NSW.

P. TENUIRIMA Tayl., London J. Bot. 3: 645. 1844. *P. tenuiscypha Tayl. ex Hook., op. cit. 6: 175. 1847. "MacQuarry River." Maiden (1898, as "P. tenuissima Tayl."); Muell.-Arg. (1883a, 1888c, 1892a); Shirley (1893a); Wilson (1887, 1889b,c). NSW, Q, V.

*P. TENUIRIMA f. ISIDIOSA Muell.-Arg., Bull. Herb. Boiss. 4: 90. 1896. Muell.-Arg. (1898). V.

*P. THAMNOIDES Kurok., J. Hattori Bot. Lab. 32: 213-214. 1969. Q.

"P. TILIACEA var. CONCENTRICA Leight." Wilson (1889a). V. Nomen nudum?

P. TINCTORUM Nyl., Flora 55: 547. 1872. P. praetervisa Muell.-Arg., op. cit. 63: 276. 1880. P. coralloidea Vain., Etud. Lich. Brésil 1: 33. 1890. Bailey (1881, 1883); Hale (1965); Hue (1890-92); Muell.-Arg. (1891a); Shirley (1889a, 1891a, 1892b, 1893a, 1894); Stirton (1899b); Turner (1905); Wilson (1889b); Zahlbruckner (1896). NSW, Q.

*P. TUMESCENS Hale & Kurokawa, Contr. U.S. Nat. Herb. 36: 147. 1964. *P. limbata f. isidiosa Muell.-Arg., Flora 70: 59. 1887. NSW.

P. URCEOLATA Eschw. in Mart. var. SOREDIIFERA Muell.-Arg., op. cit. 63: 266. 1880. Muell.-Arg. (1892a); Shirley (1889a, 1892a). NSW, Q.

P. URCEOLATA var. SUBCETRATA Muell.-Arg., op. cit. 66: 46. 1883. Muell.-Arg. (1883a, 1892a); Shirley (1889a, 1892a); Zahlbruckner (1896). Q.

P. VERSICOLOR Muell.-Arg., op. cit. 64: 506. 1881 (non Ach. 1803). P. molliuscula of Australian reports, non Ach. Bailey (1881, 1883); Kurokawa (1969); F. Mueller (1881); Tate (1882); Willis (1959b); Wilson (1887). Q, SA, V.

*P. VICTORIANA A. Zahlbr., Cat. Lich. Univ. 6: 222. 1929. P. hypoleuca Muell. Arg., Flora 70: 317. 1887 (non Tuck.). P. novae-hollandiae A. Zahlbr., Cat. Lich. Univ. 8: 562. 1932 (nomen superfl.) V.

(Parmelia)

*P. VICTORIANA f. CORALLOIDEA A. Zahlbr., Cat. Lich. Univ. 6: 222. 1929. *P. hypoleuca f. coralloidea Muell.-Arg., Flora 70: 317. 1887. V. P. novae-hollandiae f. coralloidea A. Zahlbr., Cat. Lich. Univ. 8: 562. 1932 (nomen superfl.). NSW?, V. WA.

*P. VIRENS var. SOREDIATA Muell.-Arg., Flora 69: 256. 1886. Q. Muell.-Arg. (1892a).; Hale (1965) treats this as a nomen dubium. The type is in poor condition.

*P. XANTHOMELANA Muell.-Arg., op. cit. 66: 48. 1883. Vainio (1900). V.

PARMELIELLA

*P. BAEUERLENII Muell.-Arg., Flora 69: 286. 1886. Muell.-Arg. (1887b). "Brogers Creek, E Australia."

*P. COERULESCENS Muell.-Arg., Hedwigia 32: 122. 1893. Shirley (1894). Q.

P. CORALLINOIDES (Hoffm.) A. Zahlbr., Ann. Naturhist. Hofmus. 13: 462. 1899. Stereocaulon coralloides Hoffm., Deutschl. Fl. p. 129. 1796. Pannaria tryptophylla (Ach.) Mass., Ricerch. Auton. Lich. p. 112. 1853. Lecidea tryptophylla Ach., Kongl. Svenska Vetensk.-Akad. Handl. p. 272. 1808. Bailey (1881, 1883); Shirley (1889a). Q.

*P. DIFFRACTA Muell.-Arg., Hedwigia 32: 123. 1893. V.

*P. DUPLICATA Muell.-Arg., Flora 66: 78. 1883. NSW.

P. MICROPHYLLA (Sw.) Muell.-Arg., op. cit. 72: 507. 1889. Lichen microphyllus Sw., Kongl. Svenska Vetensk.-Akad. Handl. p. 301. 1791. Muell.-Arg. (1893). V.

P. NIGROCINCTA (Mont.) Muell.-Arg., Flora 64: 86. 1881. Parmelia nigrocincta Mont., Ann. Sci. Nat., Bot. (2) 4: 91. 1835. Pannaria nigrocincta Nyl., op. cit. (4) 3: 182. 1855. Shirley (1889a); Wilson (1887, 1889c). Q, V.

P. PANNOSA (Sw.) Muell.-Arg., Flora 64: 86. 1881. Lichen pannosus Sw., Nova Gen. Sp. Pl. p. 146. 1788. *Pannaria cervina Kremp., Verh. zool.-bot. Ges. Wien 30: 339. 1880. Bailey (1881, 1883); F. Mueller (1881); Muell.-Arg. (1887c); Shirley (1889a). Q.

P. PLUMBEA (Lightf.) Muell.-Arg., Bull. Herb. Boiss. 2, Appendix I, p. 44. 1894. Lichen plumbeus Lightf., Fl. Scotica 2: 826. 1777. Coccocarpia plumbea C. Muell., Bot. Zeitung 15: 386. 1857. Bailey (1881, 1883); F. Mueller (1881); Shirley (1889a). Q.

PARMELIOPSIS

P. ALEURITES (Ach.) Nyl., Syn. Lich. 2: 54. 1863. Lichen aleurites Ach., Lichenogr. Suec. Prodrom. p. 177. 1798. Wilson (1887). V.

P. HYPEROPTA (Ach.) Vain., Medd. Soc. Fauna Flora Fenn. 6: 127. 1881. Parmelia hyperopta Ach., Syn. Lich. p. 208. 1814. Platysma diffusa Nyl., Flora 55: 247. 1872. Wilson (1887). V.

PARMENTARIA

P. ASTROIDEA Fée, Essai Crypt. Ecorc. Officin. p. xci, 70. 1824. Muell.-Arg. (1895b); Shirley (1893a). Q. [These records may refer to Bottaria cruentata which see.]

(Parmentaria)

*P. AUSTRALIENSE (Stirt.) Muell.-Arg., Nuovo Giorn. Bot. Ital. 23: 401. 1891. Plagiothelium australiense Stirt., Trans. Proc. Roy. Soc. Victoria 17: 75. 1881. Bailey (1881, 1883); F. Mueller (1881); Muell-Arg. (1895b). Q.

*P. BAILEYI Muell.-Arg., Hedwigia 30: 54. 1891. (orig. orthogr. 'baileyana'). Muell.-Arg. (1895b); Shirley (1891b). Q.

*P. GREGALIS (Knight) Muell.-Arg., Flora 70: 426. 1887. Trypethelium gregale Knight ex Bailey, Syn. Queensland Flora 1. Suppl. p. 77. 1886. Muell.-Arg. (1895b). Q.

*P. GROSSA Muell.-Arg., Rep. Austral. Assoc. Adv. Sci. p. 464. 1895. Shirley (1895). Q.

P. INTERLATENS (Nyl.) Muell.-Arg., Flora 66: 244. 1883. Astrothelium interlatens Nyl., Bull. Soc. Linn. Normand. (2) 2: 134. 1868. Muell.-Arg. (1895b); Shirley (1895). Q.

*P. MICROSPORA Muell.-Arg., Flora 70: 427. 1887. Trypethelium nanosporum Knight ex Bailey, Syn. Queensland Flora 1. Suppl. p. 78. 1886 (nomen illegit.). Muell.-Arg. (1895b). Q.

*P. PALLIDA (Knight ex Bailey) Shirley, Lich. Fl. Queensland 4: 163. 1890. Trypethelium pallidum Knight ex Bailey, Syn. Queensland Fl. 1. Suppl. p. 77. 1886. Muell.-Arg. (1895b); Shirley (1888b, 1889d). Q.

*P. PAPILLATA (Knight ex Bailey) Shirley, Lich. Fl. Queensland 4: 161. 1890. Trypethelium papillatum Knight ex Bailey, Syn. Queensland Fl. 1. Suppl. p. 76. 1886. Muell.-Arg. (1895b). Q.

*P. PLANA (Knight ex Bailey) Shirley, Lich. Fl. Queensland 4: 162. 1890. Trypethelium planum Knight ex Bailey, Syn. Queensland Fl. 1. Suppl. p. 77. 1886. Q.

P. RAVENELII (Tuck.) Muell.-Arg., Flora 68: 249. 1885. Pyrenastrum ravenelii Tuck., Amer. J. Arts Sci. (2) 25: 429. 1858. Muell-Arg. (1893a). V.

*P. SUBASTROIDEA Muell.-Arg., Rep. Austral. Assoc. Adv. Sci. p. 463. 1895. Shirley (1895). Q.

*P. SUBASTROIDEA var. SUBSIMPLEX Muell.-Arg., ibid. Q.

*P. SUBPLANA (Knight ex Bailey) Muell.-Arg., Flora 70: 426. 1887. Trypethelium subplanum Knight ex Bailey, Syn. Queensland Fl. 1 Suppl. p. 77. 1886. Muell.-Arg. (1895b). Q.

*P. SUBUMBILICATA (Knight ex Bailey) Muell.-Arg., Flora 70: 426. 1887. Trypethelium subumbilicatum Knight ex Bailey, Syn. Queensland Fl. 1 Suppl. p. 76. 1886. Muell.-Arg. (1895b); Wilson (1889a). Q, V.

*P. TOOWOOMBENSIS Muell.-Arg., Ann. K. K. Naturhist. Hofmus. 7: 305. 1892. Muell.-Arg. (1895b); Shirley (1895). Q.

PELTIGERA

P. APHTHOSA (L.) Willd., Fl. Berol. p. 347. 1787. Lichen aphthosus L., Sp. Pl. p. 1148. 1753. Peltidea aphthosa Ach., Method. Lich. p. 287. 1803. Wilson (1887); Zahlbruckner (1896). NSW, V.

P. CANINA (L.) Willd., Fl. Berol. p. 347. 1787. Lichen caninus L., Sp. Pl. p. 1149. 1753. Wilson (1887). V.

P. DOLICHORHIZA (Nyl.) Nyl., Lich. Nov. Zel. p. 43. 1888. P. polydactyla var. dolichorhiza Nyl., Syn. Lich. 1: 327. 1860. Hellbom (1896); Krempelhuber (1880); Murray (1960c); Weber (1969); Wilson (1887, 1889b). WA.

*P. DOLICHORHIZA f. PSEUDOCRISPOIDES Gyel., Acta Fauna Fl. Univ. (2) 1: 5. 1933. Q.

P. HASZLINSZKYI Gyel. ex Anders, Strauch- und Laubflecht. Mitteleurop. p. 44. 1928. Gyelnik (1932). NSW.

P. LEPTODERMA Nyl., Syn. Lich. 1: 325. 1860. Solorina sorediifera Nyl., loc. cit. p. 331. Gyelnik (1932); Wilson (1889a). NSW.

P. POLYDACTYLA (Neck.) Hoffm., Descr. Adumbr. Pl. Lich. 1: 19. 1790. Lichen polydactylus Neck., Method. Musc. p. 85. 1771. Bailey (1883); Bibby & Smith (1954); Darbishire (1912); Hampe (1853), F. Mueller (1881); Shirley (1889a); Turner (1905); Wilson (1889c); Zahlbruckner (1896). NSW, Q, V, WA.

P. POLYDACTYLA var. CONJUNGENS Muell.-Arg., Flora 66: 22. 1883. Shirley (1889b). Q.

P. POLYDACTYLA var. DISSECTA Muell.-Arg., op. cit. 74: 374. 1891. Muell.-Arg. (1892a); Shirley (1893a). Q, V.

*P. POLYDACTYLA f. PUDENS F. Wils., Victoria Naturalist 6: 179. 1890. V.

P. PULVERULENTA (Tayl.) Nyl., Syn. Lich. 1: 325. 1860. Peltidea pulverulenta Tayl., London J. Bot. 6: 184. 1847. Wilson (1887). V.

P. RUFESCENS (Weis) Humb., Fl. Friburg. Specim. p. 2. 1793. Lichen caninus, rufescens Weis, Pl. Crypt. Fl. Goettingens. p. 79. 1770. Wilson (1887). V.

P. SPURIA (Ach.) DC ex Lam. & DC., Flore Franç. ed. 3, 2: 406. 1805. Lichen spurius Ach., Lich. Suec. Prodr. p. 159. 1798. Wilson (1889b, in doubt).

*P. SUBHORIZONTALIS Gyel., Ann. Cryptog. Exot. 5: 39. 1932. NSW.

PELTULA

PELTULA EUPLOCA (Ach.) Wetmore, Ann. Missouri Bot. Gard. 57: 184. 1970. Lichen euplocus Ach., Lich. Suec. Prodr. p. 141. 1798. Heppia euploca Vainio, Acta Soc. Fauna Fl. Fenn. 49 (2): 14. 1921. Heppia guepinii (Del. in Duby) Nyl. in Hue, Rev. Bot. Bull. Mens. 5: 18. 1886. Endocarpiscium guepinii Nyl., Flora 47: 487, footnote. 1864. Hue (1890-92); Muell.-Arg. (1893); Nylander (1857). V.

PERTUSARIA

*P. ABERRANS Muell.-Arg., Bull. Herb. Boiss. 1: 42. 1892. Muell.-Arg. (1893). V.

P. ALPINA Hepp ex Ahles, De German. Pertus., Conotr. et Phlyct. p. 12. 1860. P. leioplaca var. octospora Nyl., Lich. Scand. p. 182. 1861. Muell.-Arg. (1884a, 1891a, 1893); Shirley (1892a). Q, V.

*P. AMBLYOGONA Muell.-Arg., Bull. Herb. Boiss. 3: 638. 1895. Shirley (1896). Q.

P. ANARITHMETICA Muell.-Arg., Bull. Soc. Bot. Belg. 30: 63. 1891. Muell.-Arg. (1893). V.

*P. ARENACEA Muell.-Arg., Hedwigia 34: 29. 1895. V.

P. COMMUTATA Muell.-Arg., Flora 67: 269. 1884. Shirley (1893a). Q.

*P. CONCAVA Muell.-Arg., Bull. Herb. Boiss. 3: 640. 1895. V.

*P. CONFLUENS Muell.-Arg., loc. cit. p. 638. Shirley (1896). Q.

*P. CRASSILABRA Muell.-Arg., Hedwigia 32: 126. 1893. V.

P. DEPRESSA (Fée) Mont. var. OCTOMERA Muell.-Arg., Flora 67: 289. 1884. Shirley (1893a). Q.

P. DERMATODES Nyl., Ann. Sci. Nat., Bot. (4) 11: 241. 1859. Muell.-Arg. (1884a); Shirley (1889a). Q.

*P. DIFFRACTA Muell.-Arg., Bull. Herb. Boiss. 1: 43. 1893. V.

*P. ELLIPTICA Muell.-Arg., op. cit. 3: 635. 1895. Shirley (1896). Q.

*P. ERYTHRELLA Muell.-Arg., op. cit. 1: 41. 1893. V.

*P. GIBBEROSA Muell.-Arg., Flora 65: 486. 1882. Shirley (1889a). Q.

*P. GLEBOSA Muell.-Arg., loc. cit. p. 485. Muell.-Arg. (1884a). V.

P. GLOBULIFERA Nyl., Herbar. Mus. Fenn. p. 87. 1859 (non Mass., 1855). Zahlbruckner, Cat. Lich. Univ. 5: 192. 1928, synonymizes this with P. panyrga (Ach.) Mass., but the description in Shirley (1893a) suggests that the Australian plant, at least, is incorrectly placed under that name.

*P. GRAPHIDIOIDES Muell.-Arg., Bull. Herb. Boiss 1: 42. 1893. V.

*P. IRREGULARIS Muell.-Arg., op. cit. 3: 638. 1895. Shirley (1896). Q.

P. LACTEA (L.) Arn., Verhandl. zool.-bot. Ges. Wien 22: 283. 1872. Lichen lacteus L., Mant. 1: 132. 1767. Muell.-Arg. (1891a); Shirley (1893a). Q.

*P. LEIOCARPELLA Muell.-Arg., Bull. Herb. Boiss. 3: 636. 1895. Shirley (1896). Q.

P. LEIOPLACA (Ach.) DC. in Lam. & DC., Flore Franç. ed 3, 6: 173. 1815. Porina leioplaca Ach., Kongl. Svenska Vetensk.-Akad. Nya Handl. p. 159. 1809. Bailey (1881); Hue (1890-92); Knight (1882); F. Mueller (1881); Nylander (1886); Shirley (1889a); Stirton (1899b "var. minor Schaer."). NSW, Q.

P. LEIOPLACA var. GIBBOSA Muell.-Arg., Bull. Soc. Bot. Belg. 32: 63. 1891. Muell.-Arg. (1891a); Shirley (1892a). Q.

P. LEIOPLACELLA Nyl., Bull. Soc. Linn. Normand. (2)2: 71. 1868. Bailey (1881, 1883); F. Mueller (1881); Shirley (1889a); Wilson (1890b). Q, WA.

*P. LEIOTERA Muell.-Arg., Flora 67: 285. 1884. Bailey (1891a); Muell.-Arg. (1891a,b); Shirley (1889a). Q.

*P. LEUCOSTIGMA Muell.-Arg., loc. cit. p. 462. Shirley (1889a). Q.

*P. LEUCOSTOMOIDES A. Zahlbr., Cat. Lich. Univ. 5: 172. 1928. P. leucostoma Muell.-Arg., Bull. Herb. Boiss. 3: 636. 1895 (non Mont. & v. d. Bosch, nec Mass.). Shirley (1896). Q.

*P. LEUCOTHELIA Muell.-Arg., Bull. Herb. Boiss. 3: 637. 1895. V.

*P. LEUCOXANTHA Muell.-Arg., ibid. Shirley (1896). Q.

*P. LOPHOCARPA Koerb., Abh. Schles. Ges. vaterl. Kultur 2: 34. 1862. Hue (1890-92). NSW.

*P. MACRA Muell.-Arg., Bull. Herb. Boiss. 3: 639. 1895. Shirley (1896). Q.

P. MELALEUCA (Turn. & Borr.) Duby, Bot. Gall. 2: 673. 1830. Lichen melaleucus Turn. & Borr., in Sm. & Sowerb., Engl. Bot. 35: tab. 2461. 1813. Muell.-Arg. (1884a); Shirley (1889a). Q.

P. MELALEUCA var. OCTOSPORA Muell.-Arg., Bull. Herb. Boiss. 1: 42. 1893. Muell.-Arg. (1892a). Q.

P. MELALEUCA var. TRISPORA Muell.-Arg., ibid. V.

P. MERIDIONALIS Muell.-Arg. var. XANTHOSTOMA Muell.-Arg., Flora 64: 515. 1881. Muell.-Arg. (1891a); Shirley (1891a). Q.

*P. MICROSPORELLA A. Zahlbr., Cat. Lich. Univ. 5: 180. 1928. P. microspora Muell.-Arg., Bull. Herb. Boiss. 3: 637. 1895. (non Krempelhuber, 1876). V.

*P. MINUTA Knight ex Shirley, Proc. Roy. Soc. Queensland 6: 143. 1889. Q. Bailey (1883).

*P. MOFFATIANA Muell.-Arg., Flora 66: 79. 1883. Muell.-Arg. (1884a). V

*P. MUELLERIANA A. Zahlbr., Cat. Lich. Univ. 5: 181. 1928. P. albinea Muell.-Arg., Bull. Herb. Boiss. 3: 639. 1895 (non Tuck., 1877). Shirley (1896). Q.

P. MULTIPUNCTA (Turn.) Nyl., Lich. Scand. p. 179. 1861. Variolaria multipuncta Turn., Trans. Linn. Soc. London 9: 137. 1806. Hue (1890-92); Nylander (1886); Stirton (1899b). NSW, Q.

*P. NITIDULA Muell.-Arg., Bull. Herb. Boiss. 1: 42. 1893. V.

*P. PARATROPA Muell.-Arg., op. cit. 3: 639. 1895. V.

*P. PERSULPHURATA Muell.-Arg., Nuovo Giorn. Bot. Ital. 23: 391. 1891. Bailey (1891); Shirley (1891b). Q.

P. PERTUSA (L.) Tuck., Enum. N. Am. Lich. p. 56. 1845. Lichen pertusus L., Mant. 1: 131. 1767. Pertusaria communis DC. in Lam. & DC., Flore Franç. ed. 3, 2: 320. 1805. Wilson (1887). V.

P. PERTUSELLA Muell.-Arg., Flora 67: 283. 1884. Shirley (1889a). Q.

*P. PETROPHYES Knight, Trans. Linn. Soc. London, Bot. (2) 2: 47. 1882. Hue (1890-92); Nylander (1886); Shirley (1889a). NSW.

*P. PLICATULA Muell.-Arg., Bull. Herb. Boiss. 3: 635. 1895. Shirley (1896). Q.

P. PORINELLA Nyl., Ann. Sci. Nat., Bot. (4) 19: 321. 1863. Bailey (1881, 1883); F. Mueller (1881); Shirley (1889a). Q.

P. PUSTULATA (Ach.) Duby, Bot. Gall. 2: 673. 1830. Porina pustulata Ach., Lichenogr. Univ. p. 309. 1810. Hue (1890-92); Muell.-Arg. (1884a); Tate (1887, as "punctulata"). Q, SA.

*P. PUSTULATA var. TRIMERA Muell.-Arg., Bull. Herb. Boiss. 1: 42. 1893. V.

P. QUASSIAE (Fee) Nyl., Ann. Sci. Nat., Bot. (4) 15: 45. 1861. Porina quassiae Fée, Essai Crypt. Ecorc. Officin. p. 81. 1824. Hue (1890-92). Sine loc.

*P. RHODOTROPA Muell.-Arg., Bull. Herb. Boiss. 3: 635. 1895. Shirley (1896). Q.

*P. SCHIZOSTOMELLA Muell.-Arg., l.c., p. 637. NSW.

*P. SOREDIATA Knight ex Shirley, Proc. Roy. Soc. Queensland 6: 141. 1889. Q.

*P. STRAMINEA Muell.-Arg., Bull. Herb. Boiss. 3: 638. 1895. Shirley (1896). Q.

P. SUBFLAVENS Muell.-Arg., Proc. Roy. Soc. Edinburgh 11: 461. 1882. Shirley (1889a). Q.

P. SUBLUTESCENS A. Zahlbr., Cat. Lich. Univ. 5: 216. 1928. P. lutescens Kremp., J. Mus. Godeffroy 1 (4): 103. 1874 (non Lamy 1878). Shirley (1889a). Q.

*P. SUBRIGIDA Muell.-Arg., Bull. Herb. Boiss. 3: 636. 1895. Shirley (1896). Q.

P. SUBVAGINATA Nyl., Flora 49: 290. 1866. Muell.-Arg. (1891a, 1892a); Shirley (1891a). Q.

*P. SULPHURATA Muell.-Arg., Hedwigia 32: 125. 1893. Shirley (1893a). Q.

P. TETRATHALAMIA (Fée) Nyl., Act. Soc. Scient. Fenn. 7: 448. 1863. Trypethelium tetrathalamium Fée, Essai Crypt. Ecorc. Officin. p. 69. 1824. Pertusaria leioplacoides Muell.-Arg., Flora 64: 517. 1881. Shirley (1889a). Q.

*P. TETRATHALAMIA var. PLICATULA (Muell.-Arg.) Muell.-Arg., Rev. Mycol. 9: 84. 1887. P. leioplacoides var. plicatula Muell.-Arg., Flora 67: 302. 1884. V.

*P. THIOSPODA Knight, Trans. Linn. Soc. London, Bot. (2) 2: 47. 1882. NSW. Shirley (1889a). NSW, Q.

P. TRYPETHELIIFORMIS (Nyl.) Nyl., Ann. Sci. Nat., Bot. (4) 11: 241. 1859. P. leioplaca var. trypetheliiformis Nyl., op. cit. (4) 15: 45. 1861. Muell.-Arg. (1891a); Shirley (1889a). Q.

*P. TRYPETHELIOIDES var. HARTMANNII (Muell.-Arg.) Muell.-Arg., Flora 67: 351. 1884. Pertusaria hartmannii Muell.-Arg., op. cit. 65: 485. 1882 (overlooked by Zahlbruckner, Cat. Lich. Univ. Q.

*P. UNDULATA Muell.-Arg., Hedwigia 32: 126. 1893. Q.

P. VELATA (Turn.) Nyl., Lich. Scand. p. 179. 1861. Parmelia velata Turn., Trans. Linn. Soc. London 9: 143. 1808. Pertusaria pilulifera (Pers.) Nyl., Mem. Soc. Scienc. Nat. Cherbourg 5: 116. 1857. Bailey (1883); Knight (1884c, as "Verrucaria"); Krempelhuber (1880); F. Mueller (1881); Muell.-Arg. (1891a); Shirley (1889a). Q.

*P. VIRGINEA Muell.-Arg., Flora 65: 486. 1882. P. virginica Muell.-Arg., op. cit. 67: 351. 1884 (lapsus calami). NSW.

*P. WOOLLSIANA Muell.-Arg., op. cit. 65: 485. 1882. Muell.-Arg. (1884a). NSW.

P. WULFENII DC. in Lam. & DC., Flore Franç. ed. 3, 2: 320. 1805. P. fallax var. sulphurea Hepp, Flecht. Europ. no. 680. 1860. Shirley (1891a); Wilson (1887). Q, V.

*P. XANTHOPLACA Muell.-Arg., Flora 65: 485. 1882. Muell.-Arg. (1884a); Shirley (1889a). Q.

PHAEOGRAPHINA

*P. BANKSIAE Muell.-Arg., Bull. Herb. Boiss. 1: 59. 1893. V.

P. CAESIOPRUINOSA (Fee) Muell.-Arg., Mém. Soc. Phys. Hist. Nat. Genève 29(8): 49. 1887. Arthonia caesiopruinosa Fée, Suppl. Essai Crypt. Ecorc. Officin. p. 36. 1837. Bailey (1891a); Muell.-Arg. (1891a, b); Shirley (1891a). Q.

*P. CAESIOPRUINOSA var. MONOSPORA Muell.-Arg., Bull. Herb. Boiss. 3: 322. 1895. Shirley (1896). Q.

P. CHRYSENTERA (Mont.) Muell.-Arg., Hedwigia 30: 52. 1891. Graphis chrysenteron Mont., Ann. Sci. Nat., Bot., (2) 18: 268. 1842. Bailey (1891a); Muell.-Arg. (1891a,b); Shirley (1891b). Q.

P. CONTEXTA (Pers.) Muell.-Arg. ex Balfour, Trans. Roy. Soc. Edinburgh 31: 379. 1888. Emblemia contexta Pers. in Freycin., Voy. Uranie Bot., p. 184. 1826. Bailey (1891a); Muell.-Arg. (1891b); Shirley (1891b). Q.

P. QUASSIAECOLA (Fée) Muell.-Arg., Mem. Soc. Phys. Hist. Nat. Geneve 29 (8): 47. 1887. Fissurina quassiaecola Fée, Essai Crypt. Ecorc. Officin. p. xcii. 1824. Graphina pyelodes Wils. ex Bailey, Queensland Dept. Agric. Bull. 7: 32. 1891. Shirley (1893a); Wilson (1891c). Q.

PHAEOGRAPHIS

*P. AUSTRALIENSIS Muell.-Arg., Flora 65: 504. 1882. Muell.-Arg. (1883); Shirley (1894). V.

*P. CINERASCENS Muell.-Arg., loc.cit. p. 503. Muell.-Arg. (1893). V.

*P. ELAEINA (Knight) Muell.-Arg., Bull. Herb. Boiss. 3: 321. 1895. Graphis elaeina Knight, Trans. Linn. Soc. London, Bot. 2: 41. 1882. NSW.

*P. ELUDENS (Stirt.) Shirley, Proc. Roy. Soc. Queensland 6: 197. 1889. Graphis eludens Stirt. Proc. Roy. Soc. Victoria 17: 72. 1881 (overlooked by Zahlbruckner, Cat. Lich. Univ.). Bailey (1881, 1883); F. Mueller (1881); Shirley (1889a). Q. Zahlbruckner's entry under Phaeographis eludens (Cat. Lich. Univ. 2: 371. 1923) is hopelessly confused.

*P. EXTENUATA Muell.-Arg., Bull. Herb. Boiss. 1: 57. 1893. V.

*P. INSCRIPTA Muell.-Arg., Flora 65: 504. 1882. V.

*P. INTUMESCENS Muell.-Arg., Bull. Herb. Boiss. 1: 56. 1893. V.

P. LEIOGRAMMODES (Kremp.) Muell.-Arg., Nuovo Giorn. Bot. Ital. 23: 397. 1891. Graphis leiogrammodes Kremp., Vidensk. Meddel. Naturhist. Foren. Kjoebenhavn 5: 25. 1873. Muell.-Arg. (1891a); Shirley (1891b). Q.

P. MALACODES (Nyl.) A. Zahlbr., Cat. Lich. Univ. 2: 381. 1923. Graphis malacodes Nyl., Bull. Soc. Linn. Normand. (2) 2: 116. 1868. Bailey (1881, 1883); F. Mueller (1881); Shirley (1888b, 1889a). Q.

*P. MUCRONATA (Stirt.) A. Zahlbr., Trans. Glasgow Soc. Field Naturalists 4: 95. 1875. NSW. Bailey (1881, 1883); F. Mueller (1881); Shirley (1889a). NSW, Q.

*P. PSEUDOMELANA Muell.-Arg., Bull. Herb. Boiss. 3: 321. 1895. Shirley (1896). Q.

*P. SUBCOMPULSA Muell.-Arg., Flora 65: 503. 1882. V.

*P. SUBINTRICATA (Knight) Muell.-Arg., Bull. herb. Boiss. 3: 320. 1895. Graphis subintricata Knight, Trans. Linn. Soc. London, Bot. 2: 40. 1882. NSW.

P. SUBINUSTA (Leight.) Muell.-Arg., Flora 65: 382. 1882. Graphis subinusta Leight., Trans. Linn. Soc. London, 27: 177. 1869. Hue (1890-92); Nylander (1886). NSW.

*P. SUBTRICOSA Muell.-Arg., Bull. Herb. Boiss. 3: 320. 1895. Graphis subtricosa Knight, Trans. Linn. Soc. London, Bot. 2: 40. 1882. NSW.

PHAEOTREMA

*P. CONSIMILE Muell.-Arg., Flora 70: 398. 1887. Shirley (1889a,b). Q.

*P. CRICOTUM (F. Wils.) Muell.-Arg., Hedwigia 32: 130. 1893. Ocellularia cricota F. Wils. ex Bailey, Queensland Dept. Agr. Bull. 7: 32. 1891. Shirley (1893a). Q.

*P. EXPANSUM Shirley, Proc. Roy. Soc. Queensland 6: 189. 1889. Thelotrema expansum Knight ex Bailey, Syn. Queensland Fl. II Suppl., p. 86. 1888. Shirley (1889a). Q.

*P. LACTEUM (Kremp. ex Nyl.) Muell.-Arg., Flora 70: 398. 1887. Thelotrema lacteum Kremp. ex Nyl., op. cit. 47: 269. 1867. Hue (1890-92). Sine loc.

*P. TRYPETHELIOIDES Knight ex Shirley, Proc. Roy. Soc. Queensland 6: 189. 1889. Thelotrema trypethelioides Knight ex Bailey, op. cit. 1: 151. 1884. Shirley (1889a). Q.

PHLYCTELLA

*P. WILSONII Muell.-Arg., Bull. Herb. Boiss. 1: 43. 1893. Muell.-Arg. (1893b); Shirley (1894). V.

PHYLLOPSORA

P. BREVIUSCULA (Nyl.) Muell.-Arg., Bull. Herb. Boiss. 2, appendix 1: 45. 1894. Lecidea breviuscula Nyl., Ann. Sci. Nat., Bot. (4) 19: 339. 1863. Psora breviuscula Muell.-Arg., Flora 65: 483. 1882. Muell.-Arg. (1891a); Shirley (1889a, 1891a). Q.

*P. FOLIATA (Stirt.) A. Zahlbr., Cat. Lich. Univ. 4: 397. 1927. Lecidea foliata Stirt., Trans. Proc. Roy. Soc. Victoria 17: 71. 1881. Psora foliata Muell.-Arg., Flora 65: 483. 1882. Bailey (1881, 1883); F. Mueller (1881); Shirley (1888b, 1889). Q.

*P. FOLIATA var. ATROVIRENS (Knight ex Shirley) A. Zahlbr., Cat. Lich. Univ. 4: 397. 1927. Lecidea foliata var. atrovirens Knight ex Shirley, Proc. Roy. Soc. Queensland 6: 166. 1889. Q. Muell.-Arg. (1891a). Q.

*P. FOLIATA var. SUBCORALLINA (Muell.-Arg.) A. Zahlbr., Cat. Lich. Univ. 4: 397. 1927. Psora foliata var. subcorallina Muell.-Arg., Flora 65: 483. 1882. Lecidea foliata var. subcorallina Shirley, Proc. Roy. Soc. Queensland 6: 166. 1889. Muell.-Arg. (1891a). Q.

P. PARVIFOLIA (Pers.) Muell.-Arg., Bull. Herb. Boiss. 2: 90. 1894. Lecidea parvifolia Pers. in Gaudich., Voy. Uranie, Bot., p. 192. 1826. Bailey (1881, 1883); Crombie (1880); F. Mueller (1881); Muell.-Arg. (1882b, 1891a); Shirley (1889a); Turner (1905). NSW, Q.

P. PARVIFOLIA var. FIBRILLIFERA (Nyl.) Muell.-Arg., Bull. Soc. Bot. Belg. 32: 131. 1893. Lecidea parvifolia var. fibrillifera Nyl., Ann. Sci. Nat., Bot. (4) 15: 47. 1861. Psora parvifolia var. corallina Muell.-Arg., J. Bot. (Desvaux) 7: 55. 1893. Muell.-Arg. (1891a); Shirley (1889a), 1892a). Q.

*P. PARVIFOLIA var. GRANULOSA (Muell.-Arg.) Muell.-Arg., Bot. Jahrb. 20: 264. 1894. Psora parvifolia var. granulosa Muell.-Arg., Flora 65: 327. 1882. Lecidea parvifolia var. granulosa Shirley, Proc. Roy. Soc. Queensland 6: 166. 1889. Q.

P. PARVIFOLIA var. SUBGRANULOSA (Tuck.) Muell.-Arg., Bot. Jahrb. 20: 264. 1894. Lecidea parvifolia var. subgranulosa Tuck., Proc. Amer. Acad. Arts Sci. 6: 273. 1866. Psora parvifolia var. subgranulosa Muell.-Arg., J. Linn. Soc. London, Bot. 29: 219. 1893. Shirley (1893a). Q.

*P. SUBHYALINA (Stirt.) A. Zahlbr., Cat. Lich. Univ. 4: 401. 1927. Lecidea subhyalina Stirt., Trans. Proc. Roy. Soc. Victoria 17: 77. 1881. F. Mueller (1881); Shirley (1889a). Q, V.

PHYSCIA

P. ADGLUTINATA (Flk.) Nyl., Mém. Soc. Imp. Scienc. Nat. Cherbourg 5: 107. 1857. Lecanora adglutinata Flk., Deutschl. Fl. 4: 7. 1815. Shirley (1889a,e); Wilson (1887). Q, V.

P. AEGIALITA (Ach.) Nyl., Ann. Sci. Nat., Bot. (4) 15: 43. 1861. Parmelia aegialita Ach., Method. Lich. p. 191. 1803. Physcia confluens Nyl., Mém. Soc. Imp. Scienc. Nat. Cherbourg 5: 107. 1857. Bailey (1881); Shirley (1889a); Stirton (1889b); Weber (1969); Wilson (1889a). Q, V.

P. AIPOLIA (Ehrh. in Humb.) Hampe in Fuernr., Naturh. Topogr. Regensb. 2: 249. 1839. Lichen aipolius Ehrh. in Humb., Fl. Frib. Specim., p. 19. 1793. Stirton (1899b). Q.

P. AIPOLIA var. ACRITA (Ach.) Hue, J. Bot. (Desvaux) 4: 266. 1890. Parmelia aipolia var. acrita Ach., Lichenogr. Univ. p. 477. 1810. Physcia stellaris f. acrita Nyl., Lich. Scand. p. 111. 1861. Muell.-Arg. (1892b); Shirley (1893a, 1894); Stirton (1899b). Q, V.

*P. ALBATA (F. Wils.) Hale, Bryologist 66: 73. 1963. *Parmelia albata F. Wils., Victoria Naturalist 6: 69. 1889, Pap. Proc. Roy. Soc. Tasmania for 1892, p. 173. 1893. Muell.-Arg. (1896, 1898); Wilson (1889a). V.

P. ALBICANS (Pers.) Thoms., Nova Hedwigia Beih. 7: 88. 1963. Parmelia albicans Pers., Annal. Wetterau 2: 17. 1811. Physcia crispa Nyl., Syn. Lich. 1: 423. 1860. Parmelia crispa Pers. in Gaudich., Voy. Uranie, Bot. p. 196. 1826 (non Ach.). Bailey (1883); Krempelhuber (1880); F. Mueller (1881); Shirley (1889a); Stirton (1899b). NSW, Q.

(P. ALBICANS) *P. CRISPA var. LINEARIS Muell.-Arg., Bull. Herb. Boiss. 4: 91. 1896. Muell.-Arg., (1898). V.

P. CAESIA (Hoffm.) Hampe in Fuernr., Naturh. Topogr. Regensb. 2: 250. 1839. Lichen caesius Hoffm., Enum. Lich. p. 65. 1788. *Parmelia alboplumbea Tayl., London J. Bot. 6: 161. 1847. WA. *P. propinqua Laur., Linnaea 2: 40. 1827. Sine loc. Hue (1890-92); Muell.-Arg. (1888c); Weber (1969); Wilson (1887). V, WA.

(Physcia)

P. CALLOSA Nyl., Flora 52: 119. 1869. NSW. Kosciusko State Park; on huge boulders, Spencers Creek valley between Perisher Valley and Chalet, 5500 ft., 4 Jan. 1968, Weber & McVean L-49290 (COLO).

P. CLEMENTII (Sm.) Lynge in Rabenh., Krypt.-Fl. Deutschl. 9, 6/1: 93. 1935. Lichen clementii Sm. in Sm. & Sowerby, Engl. Bot. 25: tab. 1779. 1807. Physcia astroidea (Bagl.) Nyl., Act. Soc. Linn. Bordeaux 21: 308. 1856. Anaptychia astroidea Bagl., Mem. Acad. Sci. Torino (2) 17: 385. 1857. Wilson (1889a). V.

*P. EXCELSIOR Stirt. ex Bailey, Queensland Agr. J. 5: 40. 1899. Q.

*P. GLAUCOVIRESCENS Nyl., Syn. Lich. 1: 419. 1860. Shirley (1889). Q.

*P. HAMILTONII Muell.-Arg., Flora 69: 258. 1886. NSW.

P. INTEGRATA Nyl., Syn. Lich. 1: 424. 1860. Shirley (1889a). Q.

P. INTEGRATA var. OBSESSA (Mont.) Vain., Etud. Lich. Brésil 1: 141. 1890. Parmelia obsessa Mont. in Sagra, Hist. l'Ile Cuba, Bot., p. 227. 1838-42 (non Ach.). F. Mueller (1881). Q.

*P. LACINIATULA Stirt., Trans. Proc. New Zealand Inst. 32: 82. 1899. Stirton (1899a). NSW.

*P. MELANENTA Knight, Trans. Linn. Soc. London, Bot. 2: 48. 1882. NSW.

*P. MELANOCLINA Knight, loc. cit. p. 49. NSW.

*P. NODOSA F. Wils., Victoria Naturalist 6: 61. 1889. Nomen nudum.

P. ORBICULARIS (Neck.) Poetsch, Syst. Aufz. samenlos. Pfl. p. 247. 1872. Lichen orbicularis Neck., Method. Musc. p. 38. 1771. Physcia obscura (Ehrh.) Hampe in Fuernr., Naturh. Topogr. Regensb. 2: 249. 1839. Lichen obscurus Ehrh., Pl. Crypt. Exs. No. 177. 1785. Muell.-Arg. (1892b); Nylander (1857). WA.

P. PICTA (Sw.) Nyl., Mém. Soc. Imp. Scienc. Nat. Cherbourg 3: 175. 1855. Lichen pictus Sw., Nov. Gen. Sp. Pl. p. 146. 1788. *Parmelia plumosa Tayl., London J. Bot. 6: 173. 1847. Physcia applanata (Fée) A. Zahlbr., Cat. Lich. Univ. 7: 581. 1931. Parmelia applanata Fée, Essai Crypt. Ecorc. Officin. p. 216. 1824. Physcia picta var. sorediata Muell.-Arg., Flora 62: 292. 1879. Bailey (1881, 1885); Crombie (1880); Hue (1890-92); F. Mueller (1881); Muell.-Arg. (1891a); Nylander (1857, 1886); Shirley (1889a, 1891b, 1892b); Stirton (1899b); Turner (1905); Wilson (1890b). NSW, Q, WA.

P. PICTA f. ISIDIOPHORA Nyl., Flora 50: 3. 1867. Bailey (1891a); Muell.-Arg. (1891b); Shirley (1891a). Q.

P. SETOSA (Ach.) Nyl., Syn. Lich. 1: 429. 1860. Parmelia setosa Ach., Syn. Lich. p. 203. 1814. Krempelhuber (1860); F. Mueller (1881); Weber (1969). NSW.

P. SPARSA Nyl., Syn. Lich. 1: 429. 1860. Stirton (1899b). Q.

P. STELLARIS (Ach.) Nyl., Act. Soc. Linn. Bordeaux 21: 307. 1856. Parmelia stellaris Ach., Method. Lich. p. 209. 1803. Fries (1846-47); F. Mueller (1881); Nylander (1857); Shirley (1889a, 1892b); Tate (1887); Wilson (1887, 1889b,c). NSW, Q, SA, V.

P. STELLARIS var. RADIATA (Ach.) Nyl., J. Linn. Soc. London, Bot. 9: 249. 1865. Parmelia stellaris var. radiata Ach., Lichenogr. Univ. p. 477. 1810. Stirton (1899b). Q.

(Physcia)

*P. SUBLURIDA Stirt., Trans. Proc. Roy. Soc. Victoria 17: 69. 1881. Bailey (1881, 1883); F. Mueller (1881); Shirley (1889a); Stirton (1899a). Q.

P. TENELLA (Scop.) DC. emend. Bitt., Jahrb. Wiss. Bot. 36: 431. 1901. Lichen tenellus Scop., Fl. Carniol. ed. 2, 2: 394. 1772. Wilson (1887, as "P. stellaris var. tenella Scop."). V.

*P. TRIBACIA (Ach.) Nyl. var. TENUIS Muell.-Arg., Flora 69: 257. 1886. Tate (1887). SA.

P. VIRELLA (Ach.) Flag., Rév. Mycol. 13: 110. 1891. Lichen virellus Ach., Lichenogr. Suec. Prodrom. p. 108. 1798. Muell.-Arg. (1892b). WA.

PHYSMA

P. AMPHIURIUM (Nyl.) A. Zahlbr., Cat. Lich. Univ. 3: 24. 1924. Collema amphiurium Nyl., Ann. Sci. Nat., Bot. (4) 12: 281. 1859. P. byrsinum var. amphiurium Muell.-Arg., Hedwigia 32: 47. 1891. Bailey (1891a); Shirley (1891b). Q.

P. BYRSINUM (Ach.) Muell.-Arg., Flora 68: 531. 1885. Parmelia byrsea Ach., Method. Lich. p. 222. 1803. Collema byrsinum Ach., Lichenogr. Univ. p. 642. 1810. Bailey (1881, 1883); F. Mueller (1881); Muell.-Arg. (1891a); Shirley (1888b, 1889a, 1892b); Stirton (1899b); Turner (1905a,b). NSW. Q.

*P. GLOBIFERUM Hue, Bull. Soc. Bot. France 61: 333. 1914. Q.

PILOPHORON

P. ACICULARE (Ach.) Nyl., Mém. Soc. Scienc. Nat. Cherbourg 5: 96. 1857. Baeomyces acicularis Ach., Method. Lich. p. 328. 1803. Hue (1890-92); Leighton (1867). NSW.

*P. CONGLOMERATUM F. Wils., Victoria Naturalist 6: 68. 1889, J. Linn. Soc. London, Bot. 28: 372. 1891. Wilson (1889a). V.

PLACOLECANORA

*P. AUSTRALICA Raes., Arch. Soc. Zool. Bot. Fenn. "Vanamo" 3: 178. 1949. Q.

PLACOPSIS

P. BRACHYLOBA (Muell.-Arg.) M. Lamb, Lilloa 13: 212. 1947. Placodium brachylobum Muell.-Arg., Bull. Herb. Boiss. 4: 93. 1896. Lecanora brachyloba A. Zahlbr., Cat. Lich. Univ. 5: 655. 1928. Lamb (1947); Muell.-Arg. (1898). Q.

P. GELIDA (L.) Nyl. Wilson (1887). V. This old record must apply to another species, since Lamb does not report it from Australia. Earlier, the epithet was commonly given to any Placopsis.

P. PARELLINA (Nyl.) M. Lamb, Lilloa 13: 244. 1947. Lecanora parellina Nyl., Ann. Sci. Nat., Bot. (4) 3: 157. 1855. Not attributed to Australia by Lamb (1947). NSW: Point Lookout, New England Nat. Park, 26 Oct. 1967, Weber & McVean L-49339; V: Tarra Valley, 13 Aug. 1961, S. Thrower 14; Noojee, Nov. 1964, McVean 64123 (COLO).

P. PERRUGOSA (Nyl.) Nyl., Lich. Nov. Zel. p. 57. 1888. Lecanora perrugosa Nyl., Flora 48: 338. 1865. Lamb (1947). V.

P. RHODOPHTHALMA (Muell.-Arg.) Raes., Ann. Soc. Zool. Bot. Fenn. "Vanamo" 2 (1): 25. 1932. Lecanora rhodophthalma Muell.-Arg., Flora 62: 164. 1879. Shirley (1892a). Q.

PLATISMATIA

P. GLAUCA (L.) Culb. & Culb., Contr. U.S. Natl. Herb. 34: 530. 1968. Lichen glaucus L., Sp. Pl., p. 1148. 1753. Cetraria glauca Ach., Method. Lich. p. 296. 1803. Bailey (1881, 1883); Laurer (1827); F. Mueller (1881); Shirley (1889a). Q.

PLEUROTHELIOPSIS

*P. AUSTRALIENSIS (Muell.-Arg.) A. Zahlbr., Cat. Lich. Univ. 1: 513. 1922. Pleurothelium australiense Muell.-Arg., Nuovo Giorn. Bot. Ital. 23: 401. 1891. Bailey (1891); Muell.-Arg. (1895b); Shirley (1891b). Q.

PLEUROTREMA

*P. PYRENULOIDES Muell.-Arg., Rep. Austral. Assoc. Adv. Sci. p. 462. 1895. Shirley (1895). Q.

POLYBLASTIOPSIS

*P. (?) COARCTATA (Stirt.) A. Zahlbr., Cat. Lich. Univ. 1: 348. 1922. Verrucaria coarctata Stirt. ex Bailey, Queensland Agric. J. 5: 39. 1899. Q.

P. GEMINELLA (Trev.) A. Zahlbr. in Engl. Prantl, Nat. Pflanzenfam. I Teil, Abt. I*, p. 65. 1903. Verrucaria geminella Nyl., Flora 61: 381. 1858. Polyblastia geminella Trev., Conspect. Verruc., p. 14. 1860; Muell.-Arg., Flora 65: 401. 1882. Muell.-Arg. (1895b); Shirley (1895). Q.

*P. GREGANTULA (Muell.-Arg.) A. Zahlbr., Cat. Lich. Univ. 1: 349. 1922. Polyblastia gregantula Muell.-Arg., Rep. Austr. Assoc. Adv. Sci. p. 454. 1895. Shirley (1895). Q.

*P. NUDATA (Muell.-Arg.) A. Zahlbr., Cat. Lich. Univ. 1: 351. 1922. Polyblastia nudata Muell.-Arg., Hedwigia 32: 135. 1893. Muell.-Arg. (1895b); Shirley (1893a). Q.

*P. PERTUSARIOIDEA (Kremp.) A. Zahlbr. Cat. Lich. Univ. 1: 351. 1922. Pyrenula pertusarioidea Kremp., Verh. zool.-bot. Ges. Wien 30: 342. 1880. Polyblastia pertusarioidea Muell.-Arg., Flora 70: 80. 1887. F. Mueller (1881); Muell.-Arg. (1895b). Q.

*P. (?) TICHOSPORA (Knight) A. Zahlbr., Cat. Lich. Univ. 1: 352. 1922. [as "trichospora"]. Verrucaria tichospora Knight, Trans. Linn. Soc. London, Bot. 2: 40. 1882. NSW. Polyblastia tichospora Shirley, Lich. Fl. Queensland 4: 180. 1890. NSW, Q.

*P. VELATA (Muell.-Arg.,) A. Zahlbr., Cat. Lich. Univ. 1: 353. 1922. Polyblastia velata Muell.-Arg., Flora 70: 428. 1887. Muell.-Arg. (1895b); Shirley (1889b). Q.

PORINA

P. AFRICANA Muell.-Arg., Linnaea 63: 41. 1880. Porina limitata Knight ex Bailey, Syn. Queensland Fl. 1 Suppl. p. 73. 1886. Shirley (1895). Q.

*P. ARAUCARIAE Muell.-Arg., Nuovo Giorn. Bot. Ital. 23: 402. 1891. Bailey (1891); Muell.-Arg., (1895b); Shirley (1891b). Q.

*P. BACILLIFERA Muell.-Arg., Flora 65: 517. 1882. Bailey (1891a); Muell.-Arg. (1891b, 1895b); Shirley (1890). Q.

*P. BELLENDENICA Muell.-Arg., Hedwigia 30: 56. 1891. Bailey (1891a); Muell.-Arg. (1895b); Shirley (1891b). Zahlbruckner, Cat. Lich. Univ. 1: 367. 1922, lists this incorrectly as P. bellendica. Q.

P. CONICA R. Sant., Symb. Bot. Upsal. 12(1): 232. 1952. Q.

*P. CORRUGATA Muell.-Arg., Bull. Herb. Boiss. 1: 63. 1893. V.

*P. ELEGANTULA Muell.-Arg., ibid. V.

*P. ENTEROXANTHA Knight ex Bailey, Syn. Queensland Fl. 1. Suppl. p. 72. 1886. Q.

P. EPIPHYLLA var. ATRICEPS Vain. London J. Bot. 34: 295. 1896. Santesson (1952). Q.

P. EPIPHYLLA (Fée) Fée, Essai Crypt. Ecorc. Officin., Suppl. p. 76. 1837. Porina americana Fée var. epiphylla Fée, Dictionn. Class. Hist. Nat. 17, p. 26. 1831. Phylloporina epiphylla Muell.-Arg., Lich. epiphylli novi, p. 21. 1890. Muell.-Arg. (1895b); Santesson (1952); Shirley (1895). Q.

*P. EXASPERATA Knight ex Bailey, Syn. Queensland Fl., 1. Suppl., p. 73. 1886. Muell.-Arg. (1895b); Shirley (1890). Q.

*P. FULVULA Muell.-Arg., Bull. Herb. Boiss. 3: 325. 1895. Mueller Arg. (1895c); Shirley (1896). Q.

P. GLAUCA Muell.-Arg., Bot. Jahrb. 6: 399. 1885. Muell.-Arg. (1895b); Shirley (1895). Q.

*P. IMPRESSA R. Sant., Symb. Bot. Upsal. 12(1): 219. 1952. Q.

P. INTERNIGRANS (Nyl.) Muell.-Arg., Rep. Austral. Assoc. Adv. Sci. p. 452. 1895. Verrucaria mastoidea var. internigrans Nyl., Bull. Soc. Linn. Normand. (2) 2: 123. 1868. Shirley (1895). Q.

P. LIMBULATA (Kremp.) Vain., Ann. Acad. Sci. Fenn. ser. A, 15: 363. 1921. Verrucaria limbulata Kremp., Lich. foliic. Beccari Ins. Borneo, p. 17. 1874 (as "limbolata"). Santesson (1952). Q.

*P. LIMITATA Knight ex Shirley, Lich. Fl. Queensland 4: 169. 1890. Q.

P. LUCIDA R. Sant., Symb. Bot. Upsal. 12(1): 240. 1952. Q.

P. MASTOIDEA (Ach.) Muell.-Arg., Bot. Jahrb. 6: 399. 1885. Pyrenula mastoidea Ach., Syn. Lich. p. 122. 1814. Verrucaria "maestroides" Ach. of reports. Bailey (1881, 1883); F. Mueller (1881); Muell.-Arg. (1891a, 1895b). Q.

P. MICROMMA (Mont.) Mass., Fragment. Lich. p. 26. 1855. Verrucaria micromma Mont., Ann. Sci. Nat., Bot. (2) 19: 57. 1843. Muell.-Arg. (1895b). Q.

P. MULTIPUNCTATA R. Sant., Symb. Bot. Upsal. 12(1): 216. 1952. Q.

P. NANA Fée, Suppl. Essai Crypt. Ecorc. Officin. p. 75. 1837. Verrucaria nana Nyl., Mem. Soc. Scienc. Nat. Cherbourg 5: 138. 1857. Bailey (1881, 1883); F. Mueller (1881). Q.

P. NUCULA Ach., Syn. Lich. p. 112. 1814. Porina mastoidea Fée, Essai Crypt. Ecorc. Officin. p. 82. 1824 (non Ach.). Shirley (1889b). Q.

P. NUCULA Ach. var. ENDOCHRYSA (Mont.) A. Zahlbr., Cat. Lich. Univ. 1: 397. 1922. P. endochrysa Mont., Ann. Sci. Nat., Bot. (2) 19: 79. 1843. Q.

P. OBDUCTA (Muell.-Arg.) Schilling, Hedwigia 67: 275. 1927. Phylloporina obducta Muell.-Arg., Flora 73: 198. 1890. Santesson (1952). Q.

*P. PALLIDA Muell.-Arg., Bull. Herb. Boiss. 3: 326. 1895. Shirley (1896). Q.

*P. PERSIMILIS Muell.-Arg., Flora 70: 428. 1887. Muell.-Arg. (1895b); Shirley (1889b, as "similis;" 1890). Q.

*P. PHAEOPHTHALMA Shirley, Lich. Fl. Queensland 4: 171. 1890. *P. brisbanensis Muell.-Arg., Nuovo Giorn. Bot. Ital. 23: 402. 1891. Q. Muell.-Arg. (1895b); Shirley (1893a).

*P. PLATYSTOMA Muell. Arg., Bull Herb. Boiss. 3: 326. 1895. V.

*P. PRAESTANTIOR Muell.-Arg. var. NANA Shirley, Lich. Fl. Queensland 4: 170. 1890. Muell.-Arg. (1895b). Q.

*P. RHAPHIDOSPORA (Knight) Muell.-Arg., Rep. Austral. Assoc. Adv. Sci. p. 453. 1895. Verrucaria rhaphidospora Knight, Trans. Linn. Soc. London, Bot. 2: 40. 1882. Arthopyrenia rhaphidospora Shirley, Lich. Fl. Queensland 4: 175. 1890. Hue (1890-92); Nylander (1886); Shirley (1895). NSW, Q.

*P. RUDIS (Muell.-Arg.) Muell.-Arg., Ann. K.K. Naturhist. Hofmus. 7: 305. 1892. Porina mastoidea Fée var. rudis Muell.-Arg., Flora 65: 517. 1882. Muell.-Arg. (1895b); Shirley (1890, 1895). Q.

P. RUFULA (Kremp.) Vain., Acta Soc. Fauna Fl. Fenn. 7: 227. 1890. Verrucaria rufula Kremp., Lich. foliic. Beccari Ins. Borneo, p. 20. 1874. Santesson (1952). NSW.

P. SEMECARPI Vain., Ann. Acad. Sci. Fenn. ser. A, 15: 367. 1921. Santesson (1952). NSW, Q.

*P. SUBARGILLACEA Muell.-Arg., Bull. Herb. Boiss. 1: 64. 1893. Muell.-Arg. (1895b). Q.

*P. SUBARGILLACEA var. NIGRATA Muell.-Arg., Rep. Austral. Assoc. Adv. Sci. p. 453. 1895. Q.

P. TETRACERAE (Ach.) Muell.-Arg., Bot. Jahrb. 6: 401. 1885. Verrucaria tetracerae Ach., Method. Lich. p. 121. 1803. Muell.-Arg. (1895b); Shirley (1895). Q.

P. VARIEGATA Fée, Suppl. Essai Cryptog. Ecorc. Officin. p. 75. 1837. Muell.-Arg. (1895b); Shirley (1895). Q.

P. VIRESCENS (Kremp.) Muell.-Arg., Flora 66: 331. 1883. Verrucaria virescens Kremp., Nuovo Giorn. Bot. Ital. 7: 53. 1875. Santesson (1952). Q.

*P. WILSONIANA Muell.-Arg., Bull. Herb. Boiss. 1: 63. 1893. V.

POROCYPHUS

*P. LICHINELLOIDES A. Henssen, Symb. Bot. Upsal. 18(1): 68. 1963. WA.

PSEUDOCYPHELLARIA

P. ARGYRACEA (Del.) Vain., Hedwigia 37: 34. 1898. Sticta argyracea Del., Hist. Lich. Sticta p. 91. 1822. Nylander (1857); Weber (1969). NSW.

(STICTA) ARGYRACEA f. ISIDIOSA (Muell.-Arg.) A. Zahlbr., Cat. Lich. Univ. 3: 371. 1925. Stictina argyracea f. isidiosa Muell.-Arg., Hedwigia 30: 48. 1891. Bailey (1891a); Shirley (1891b). Q.

P. ARGYRACEA var. SOREDIIFERA (Del.) Malme, Bih. Kongl. Svenska Vetensk. Akad. Handl. 25 (III, no. 6): 24. 1899. Sticta argyracea var. sorediifera Del., Hist. Lich. Sticta p. 92. 1822. Stizenberger (1895). Sine loc.

P. AURATA (Ach.) Vain., Etud. Lich. Brésil 1: 183. 1890. Sticta aurata Ach., Method. Lich. p. 277. 1803. Lobaria aurata O. Kuntze, Revis. Gen. Pl. 2: 876. 1891. Bailey (1881, 1883); Bibby (1954); Hampe (1853); Hellbom (1896); Hue (1890-92); Krempelhuber (1880); F. Mueller (1881); Muell.-Arg. (1891a); Nylander (1857); Shirley (1892b); Turner (1905); Wilson (1888, 1889c); Zahlbruckner (1896). NSW, Q, V.

(STICTA) AURATA var. ANGUSTATA Mont. in Gaud., Voyage Bonité p. 142. 1844-6. Sticta angustata Nyl. Ann. Sci. Nat., Bot. (4) 11: 254. 1859. Cheel (1912-14). Q.

*(STICTA) AURATA var. MICROPHYLLA Muell.-Arg., Flora 65: 304. 1882. Q. Cheel (1912-14); Shirley (1889b). NSW, Q, V.

*(STICTA) AURATA var.PALLIDOGLAUCESCENS Knight in Shirley, Proc. Roy. Soc. Queensland 6: 31. 1889. Q.

*P. AUSTRALIENSIS H. Magn., Acta Horti Gothob. 14: 9. 1940. WA. Weber (1971). NSW, SA, V, WA.

P. BILLARDIERI (Del.) Raes., Ann. Soc. Zool.-Bot. Fenn. "Vanamo" 2: 39. 1932. Sticta billardieri Del., Hist. Lich. Sticta p. 99. 1822. Cheel (1912-14); Shirley (1889a). NSW, Q, V.

(STICTA) BILLARDIERI var. LACINULATA Muell.-Arg., Bull. Herb. Boiss. 2, appendix 1: 36. 1894. Sticta fossulata f. lacinulata Kremp., Reise Oesterr. Freg. Novara, Bot. 1: 120. 1870. Cheel (1912-14). V.

P. CARPOLOMA (Del.) Vain., Hedwigia 37: 34. 1898. Sticta carpoloma Del., Hist. Lich. Sticta p. 159. 1822. Darbishire (1912); Hellbom (1896); Krempelhuber (1880); Magnusson (1940); F. Mueller (1881); Muell.-Arg. (1887c); Stizenberger (1895); Wilson (1887); Zahlbruckner (1896). NSW, V.

P. CINNAMOMEA (Rich.) Vain., Philipp. J. Sci. C, 8: 120. 1913. Sticta cinnamomea Rich., Voy. Découvert. l'Astrolabe, Bot. 1: 28. 1832. Sticta fragillima var. dissimilis Kremp., Reise Oesterr. Fregatte Novara, Bot. 1: 119. 1870. Stictina dissimilis Linds., Trans. Linn. Soc. London 25: 506. 1866. Stictina cinnamomea Muell.-Arg., Flora 66: 22. 1883. Hue (1890-92); Krempelhuber (1868, 1880); F. Mueller (1881); Muell.-Arg. (1883a, 1887c); Shirley (1889a); Stizenberger (1895); Wilson (1887) Q, V.

P. COLENSOI (Bab. ex Hook.) Vain., Résult. Voy. S.Y. Belgica, Bot. p. 28. 1903. Sticta colensoi Bab. ex Hook., Fl. Nov.-Zel. 2: 274. 1855. Sticta urvillei var. colensoi Nyl., Syn. Lich. 1: 360. 1860. Cheel (1912-14); Krempelhuber (1880); Stizenberger (1895); Watts (1903). NSW, V.

P. CROCATA (L.) Vain., Hedwigia 37: 34. 1898. Lichen crocatus L., Mant. Alter. p. 310. 1771. Sticta crocata Ach., Method. Lich. p. 277. 1803. Stictina crocata Nyl., Syn. Lich. 1: 338. 1860. Bailey (1883); Bibby & Smith (1954); Cheel (1912-14); Darbishire (1912); Fries (1846-47); Hampe (1853); Hellbom (1896); Hue (1890-92); Krempelhuber (1880); F. Mueller (1881); Nylander (1857); Smith (1962); Stizenberger (1895); Tate (1882); Wilson (1887); Zahlbruckner (1896). NSW, Q, SA, V.

*P. CROCATA f. CORALLOIDEA H. Magn., Acta Horti Gothob. 14: 15. 1940. V.

P. CROCATA f. ESOREDIATA (Muell.-Arg.) Vain., Résult. Voyage S.Y. Belgica, Bot. p. 29. 1903. Stictina crocata f. esorediata Muell.-Arg., Flora 66: 354. 1883. Sticta crocata f. esorediata A. Zahlbr., Cat. Lich. Univ. 3: 377. 1925. Cheel (1912-14); Muell.-Arg. (1887c, 1891a); Shirley (1889a, 1891a, 1893a, 1894); Stizenberger (1895). Q,V.

*P. CROCATA var. ISIDIOTYLA H. Magn., Acta Horti Gothob. 14: 14. 1940. NSW.

P. DISSIMULATA (Nyl.) Vain., Philipp. J. Sci. C, 8: 118. 1913. Sticta dissimulata Nyl., Syn. Lich. 1: 362. 1860. Krempelhuber (1880); F. Mueller (1881); Muell.-Arg., (1887c); Shirley (1889a); Stizenberger (1895); Zahlbruckner (1896). NSW, Q.

P. DURVILLEI (Del.) Vain., Hedwigia 38: 187. 1899. Sticta durvillei Del., Hist. Lich. Sticta, p. 170. 1882. Sticta urvillei Bab. ex Hook., Fl. Nov.-Zel. 2: 275. 1855. Sticta endochrysea var. urvillei Muell.-Arg., Mission Scient. Cap Horn, Bot. p. 157. 1888. Bailey (1883); Darbishire (1912); Krempelhuber (1880); F. Mueller (1881); Shirley (1889a, 1893a). Q.

P. FAVEOLATA (Del.) Malme, Bih. Kongl. Svenska Vetensk. Akad. Handl. 25 (III, no 6): 233. 1899. Sticta faveolata Del., Hist. Lich. Sticta p. 101. 1822. Nylander (1857); Stizenberger (1895). Sine loc.

P. FAVEOLATA var. CERVICORNIS (Nyl.) A. Zahlbr., Cat. Lich. Univ. 3: 380. 1925. "Stictina cervicornis Flot." of Australian reports. Krempelhuber (1880); F. Mueller (1881); Muell.-Arg. (1887c). V.

P. FLAVICANS (Hook. f. & Tayl.) Vain., Philipp. J. Sci. C, 8: 115. 1913. Sticta flavicans Hook. f. & Tayl., London J. Bot. 3: 648. 1844. Sticta endochrysea var. flavicans Muell.-Arg., Flora 71: 136. 1888. Cheel (1912-14); Krempelhuber (1880); Muell.-Arg. (1891a); Shirley (1889a, 1893a). NSW, Q.

(STICTA) FLAVISSIMA Muell.-Arg., Flora 66: 21. 1883. Shirley (1889a). Q.

*(STICTA) FLAVISSIMA var. SIMULANS Muell.-Arg., Bull. Herb. Boiss. 4: 89. 1896. Muell.-Arg. (1898). NSW, Q.

P. FOSSULATA (Laur.) Malme, Bih. Kongl. Svenska Vetensk.-Akad. Handl. 25 (III, no. 6): 22: 1899. *Sticta flotowiana Laur., Linnaea 2: 40. 1827. Sticta fossulata Duf. ex Nyl., Syn. Lich. 1: 363. 1860 (flotowiana seems to be the oldest specific epithet, but the combination in Pseudocyphellaria has not been made). Hellbom (1896); Hue (1890-92); F. Mueller (1881); Muell.-Arg. (1887c); Stizenberger (1895). Sine loc.

(STICTA) FRAGILLIMA Bab. ex Hook., Flora Nov. Zel. 2: 279. 1855. Stictina fragillima Nyl., Syn. Lich. 1: 335. 1860. Bailey (1881, 1883); Krempelhuber (1870); Turner (1905). NSW, Q.

*(STICTA) FRAGILLIMA var. DISSECTA (Muell.-Arg.) Hellb., Bih. Kongl. Svenska Vetensk.-Akad. Handl. 21 (III, No. 13): 32. 1896. *Stictina fragillima var. dissecta Muell.-Arg., Flora 66: 22. 1883. NSW. "Stictina fragilissima Nyl." of reports. Bailey (1891a); Krempelhuber (1880); Shirley (1889a); F. Mueller (1881); Muell-Arg. (1887c, 1891b). Q, V.

*(STICTA) FRAGILLIMA f. LINEARIS (Muell.-Arg.) A. Zahlbr., Cat. Lich. Univ. 3: 382. 1925. Stictina fragillima var. linearis Muell.-Arg., Flora 71: 23. 1888. Q. Bailey (1891a); Muell.-Arg. (1891b); Shirley (1889b); Stizenberger (1895). Q.

(STICTA) FRAGILLIMA f. PUNCTILLARIS (Muell.-Arg.) A. Zahlbr., Cat. Lich. Univ. 3: 382. 1925. Stictina punctillaris Muell.-Arg., Hedwigia 30: 48. 1891. Bailey (1891a); Shirley (1891b); Stizenberger (1895). Q.

P. FREYCINETII (Del.) Malme, Bih. Kongl. Svenska Vetensk.-Akad. Handl. 25 (III, no. 6): 34. 1899. Sticta freycinetii Del., Hist. Lich. Sticta p. 124. 1822. Krempelhuber (1868, 1880); F. Mueller (1881); Nylander (1857); Shirley (1889a); Stizenberger (1895); Zahlbruckner (1896). NSW, Q, V.

*(STICTA) FREYCINETII var. CONJUNGENS Muell.-Arg., Flora 66: 24. 1883. Stizenberger (1895). V.

P. FREYCINETII var. CHLOROLEUCA Vain., Hedwigia 38: 187. 1899. Sticta freycinetii var. isidioloma Nyl., Bull. Soc. Linn. Normand. (2) 2: 504. 1868. *Sticta freycinetii var. prolifera Muell.-Arg., Flora 66: 24. 1883. V. Hellbom (1896); Muell.-Arg. (1883a); Shirley (1893a); Stizenberger (1895). Q, V.

*(STICTA) FREYCINETTI var. TENUIS Muell.-Arg., Flora 66: 24. 1883. V.

P. GILVA (Ach.) Malme, Bih. Kongl. Svenska Vetensk.-Akad. Handl. 25 (III, no. 6): 32. 1899. Sticta gilva Ach., Method. Lich. p. 278. 1803. Lichen gilvus Ach., Lich. Suec. Prodr. p. 157. 1798. Bailey (1881, 1883); Cheel (1912-14); Darbishire (1912); Hue (1890-92); Magnusson (1940); F. Mueller (1881); Shirley (1889a); Stizenberger (1895); Wilson (1887). NSW, Q, V, WA.

P. GRANULATA (Hook. f.) Malme, Bih. Kongl. Svenska Vetensk.-Akad. Handl. 25 (III, 5): 21. 1899. Sticta granulata Hook. f., Fl. N.Z. 2: 281. 1855 (non Mont. 1855). Cheel (1912-14). NSW.

P. IMPRESSA (Hook. f. & Tayl.) Vain., Hedwigia 38: 187. 1899. Sticta impressa Hook. f. & Tayl., London J. Bot. 3: 648. 1844. Shirley (1893a); Zahlbruckner (1896). NSW, Q.

*P. KARSTENII (Muell.-Arg.) Szat., Ann. Mus. Nat. Hung., n.s. 7: 40. 1956. Sticta karstenii Muell.-Arg., Flora 64: 505. 1881. Q. Bailey (1891a); Muell.-Arg. (1891b); Stizenberger (1895); Turner (1905). NSW, Q.

*(STICTA) KARSTENII var. LINEARIS Muell.-Arg., Flora 69: 254. 1886. Bailey (1891a); Muell.-Arg. (1891b). Q.

(STICTA) LACTUCAEFOLIA (Pers.) Nyl., Lich. Nov. Zel. p. 39. 1888. Parmelia lactucaefolia Pers. in Gaudich., Voy. Uranie, Bot. p. 199. 1826. Sticta freycinetii var. fulvocinerea (Mont.) Nyl., Mém. Soc. Scienc. Nat. Cherbourg 5: 103. 1859. Pseudocyphellaria freycinetii var. lactucaefolia Malme, Bih. Kongl. Svenska Vetensk.-Akad. Handl. 25 (III, no. 6): 34. 1899. Wilson (1887). V.

(STICTA) MONTAGNEI Bab. ex Hook., Fl. Nov. Zel. 2: 284. 1855. Stictina montagneana Shirley, Proc. Roy. Soc. Queensland 6: 21. 1889. Q.

P. MOUGEOTIANA (Del.) Vain., Hedwigia 37: 36. 1898. Sticta mougeotiana Del., Hist. Lich. Sticta p. 62. 1822. Cheel (1912-14); Hue (1890-92); Stizenberger (1895). V.

P. MOUGEOTIANA f. AURIGERA (Del.) M. Lamb, Ann. Rep. Natl. Mus. Canada, 1952-53, Bull. 132: 250. 1954. Sticta mougeotiana var. aurigera Nyl., Ann. Sci. Nat., Bot. (4) 11: 254. 1859. Cheel (1912-14); Shirley (1894). NSW, Q, V.

P. MOUGEOTIANA var. DISSECTA Vain., Hedwigia 37: 34. 1898. Stictina mougeotiana var. dissecta Muell.-Arg., Bull. Soc. Bot. Belg. 31: 26. 1892. Cheel (1912-14); Magnusson (1940). NSW, Q.

*P. MOUGEOTIANA f. ISIDIOSA (Muell.-Arg.) H. Magn., Acta Horti Gothob. 14: 20. 1940. Stictina mougeotiana f. isidiosa Muell.-Arg., Bull. Herb. Boiss. 4: 89. 1896. Cheel (1912-14); Muell.-Arg. (1898). Q, V.

(STICTA) MOUGEOTIANA var. XANTHOLOMA Del., Hist. Lich. Sticta p. 63. 1822. Cheel (1912-14). NSW, V.

*P. NEGLECTA (Muell.-Arg.) H. Magn., Acta Horti Gothob. 14: 30. 1940. Stictina neglecta Muell.-Arg., Flora 70: 58. 1887. Cheel (1912-14); Stizenberger (1895). NSW, V.

P. ORYGMAEA (Ach.) Malme, Bih. Kongl. Svenska Vetensk.-Akad. Handl. 25 (III, no. 6): 28. 1899. Sticta orygmaea Ach., Method. Lich. p. 278. 1803. Hampe (1853); Krempelhuber (1880); Muell.-Arg. (1881). V.

P. PHYSCIOSPORA (Hook.) Malme, Bih. Kongl. Svenska Vetensk.-Akad. Handl. 25 (III, no. 6): 20. 1899. Sticta physciospora (Hook.) Nyl., Syn. Lich. 1: 364. 1860. Sticta fossulata var. physciospora Hook., Handb. New Zealand Fl. p. 569. 1867. Hellbom (1896); Hue (1890-92); Shirley (1889a); Wilson (1887). Q.

P. POLYSCHISTA (Mey. & Flot.) A. Zahlbr., Cat. Lich. Univ. 3: 360. 1925. Sticta variabilis var. polyschista Mey. & Flot., Nova Actorum Acad. Caes. Leop.-Carol. Nat. Cur. 19, Suppl. p. 214. 1843. Sticta subvariabilis Nyl., Flora 50: 439. 1867. Shirley (1889a); Turner (1905). NSW, Q.

P. RICHARDII (Mont.) Raes., Ann. Soc. zool.-bot. "Vanamo" 2: 39. 1932. Sticta richardii Mont., Ann. Sci. Nat., Bot. (2) 4: 89. 1835. Muell.-Arg. (1883a); Nylander (1857). Sine loc.

(STICTA) RUBELLA Hook. f. & Tayl., London J. Bot. 3: 649. 1844. Cheel (1912-14); Wilson (1889a). V.

(STICTA) SULPHUREA Schaer. in Moritzi, System. Verzeichn. p. 127. 1845-46. Sticta quercifolia Tayl. ex Hook. f., London J. Bot. 6: 177. 1847. Pseudocyphellaria quercifolia Vain., Philipp. J. Sci. C, 8: 117. 1913. Bibby (1954); Stizenberger (1895). Q.

P. THOUARSII (Del.) Degel., Acta Phytogeogr. Suec. 7: 150. 1935. Pseudocyphellaria intricata f. thouarsii (Del.) Vain., Hedwigia 37: 34. 1898. Sticta thouarsii Del., Hist. Lich. Sticta p. 90. 1822. Shirley (1894). V.

P. THOUARSII var. INTRICATA (Del.) Degel., Kongl. Goetheborgska Vetensk. Samhaellets Handl., Vetensk. Afd., ser. 6B, I, no. 7: 22. 1941. Sticta intricata Del., Hist. Lich. Sticta, p. 96. 1822. Bailey (1883); Krempelhuber (1880); F. Mueller (1881); Muell.-Arg. (1887c); Shirley (1889a). Q.

(STICTA) XANTHOSTICTA Pers. in Gaudich., Voy. Uranie, Bot. p. 201. 1826. Sticta lutescens Tayl. ex Hook. f., London J. Bot. 6: 179. 1847. Bailey (1881, 1883); Shirley (1889a). Q.

PSEUDOPYRENULA

P. CERATINA (Fée) Muell.-Arg., Mém. Soc. Phys. Hist. Nat. Genève 30(3): 29. 1888. Pyrenula ceratina Fée, Suppl. Essai Cryptog. Ecorc. Officin. p. 77. 1837. Verrucaria nitidiuscula Nyl., Acta Soc. Fenn. 7: 491. 1863. Pseudopyrenula nitidiuscula Shirley, Lich. Fl. Queensland 4: 175. 1890. Bailey (1881, 1883); F. Mueller (1881). Q.

*P. SULPHURESCENS (Muell.-Arg.) Muell.-Arg., Flora 66: 249. 1883. Arthopyrenia sulphurescens Muell.-Arg., op. cit. 65: 518. 1882. Muell.-Arg. (1895b); Shirley (1890). Q.

PSOROMA

P. BIATORINUM F. Wils., Victoria Naturalist 6: 61. 1889, nomen nudum.

*P. CAESIUM Muell.-Arg., Flora 69: 287. 1886. Muell.-Arg. (1892a); Shirley (1889b). Q.

P. CONTORTULUM F. Wils., Victoria Naturalist 6: 61. 1889, nomen nudum. = P. sphinctrinum fide Shirley (1894).

(Psoroma)

*P. CRAWFORDII Muell.-Arg., Flora 69: 287. 1886. SA.

*P. DISPERSUM Stirt., Trans. Proc. Roy. Soc. Queensland 17: 76. 1881. V. F. Mueller (1881); Shirley (1889a). Q, V.

*P. HYPNORUM (Hoffm.) S. Gray, Nat. Arr. Brit. Pl. 1: 445. 1821. Psora hypnorum Hoffm., Deutschl. Fl. p. 166. 1796. Wilson (1890a). V.

*P. KARSTENII Muell.-Arg., Flora 69: 287. 1886. Shirley (1889a). Q.

P. LINEARE F. Wils., Victoria Naturalist 6: 61. 1889, nomen nudum.

P. PALEACEUM (Fr.) Nyl., Ann. Sci. Nat., Bot. (4) 12: 293. 1859. Parmelia paleacea Fr., Lichenogr. Europ. Reform., p. 97. 1831. Psoroma hypnorum var. paleaceum Rostr., Bot. Tidsskr. 4: 96. 1871. Wilson (1890a). V.

P. PHOLIDOTUM (Mont.) Muell.-Arg., Flora 71: 45. 1888 (pr. p.) Parmelia pholidota Mont., Ann. Sci. Nat., Bot. (2) 4: 91. 1835. Pannaria pholidota Nyl., op. cit. (4) 3: 182. 1855. Bailey (1883); Krempelhuber (1880); F. Mueller (1881); Muell.-Arg. (1887c); Shirley (1889a). NSW, Q.

P. SPHINCTRINUM (Mont.) Nyl., loc. cit. p. 181. Parmelia sphinctrina Mont., op. cit. (2) 4: 90. 1835. Muell.-Arg. (1892a); Nylander (1857); Shirley (1889a, 1892b, 1894); Weber (1969); Willis (1953); Wilson (1887). Q, V, WA.

PYRENOPSIDIUM

*P. DECORTICANS Muell.-Arg., Hedwigia 31: 191. 1892. WA.

PYRENOPSIS

*P. AUSTRALIENSIS Muell.-Arg., op. cit. 32: 120. 1893. V.

PYRENULA

P. ADACTA Fée, Essai Crypt. Ecorc. Officin. p. 74. 1824. Muell.-Arg. (1891a, 1895b); Shirley (1891a). Q.

*P. ADACTA var. CINERASCENS (Muell.-Arg.) A. Zahlbr., Cat. Lich. Univ. 1: 422. 1922. P. pulchella var. cinerascens Muell.-Arg., Rep. Austral. Assoc. Adv. Sci. p. 458. 1895. Shirley (1895). Q.

*P. ANNULATA Muell.-Arg., Bull. Herb. Boiss. 1: 64. 1893. V.

P. ASPISTEA (Ach.) Ach., Ges. Naturf. Fr. Berlin. Mag. 6: 17. 1814. Verrucaria aspistea Ach., Method. Lich. p. 121. 1803. Pyrenula bonplandiae Fée, Essai Crypt. Ecorc. Officin. p. 74. 1824. Muell.-Arg. (1895b); Shirley (1895). Q.

*P. BAILEYI (Knight ex Bailey) Shirley, Lich. Fl. Queensland 4: 179. 1890. Verrucaria baileyi Kn. ex Bailey, Proc. Roy. Soc. Queensland 1: 91. 1884. Muell.-Arg. (1895). Q.

*P. BICUSPIDATA Muell.-Arg., Hedwigia 32: 136. 1893. Muell.-Arg., (1895b); Shirley (1893a). Q.

P. CERINA Eschw., Syst. Lich. p. 25. 1824. P. aurantiaca Fée, Suppl. Essai Crypt. Ecorc. Officin. p. 82. 1837. Bailey (1881, 1883); F. Mueller (1881). Q.

P. CIRCUMRUBENS (Nyl.) B. de Lesd., Bull. Soc. Bot. France 57: 463. 1910. Verrucaria circumrubens Nyl., Bull. Soc. Linn. Normand. (2) 2: 128. 1868. Bailey (1881, 1883); F. Mueller (1881); Muell.-Arg. (1895b). Q.

*P. CIRCUMRUBENS var. RUBROTECTA (Stirt.) Shirley, Lich. Fl. Queensland 4: 179. 1890. Verrucaria circumrubens var. rubrotecta Stirt., Trans. Proc. Roy. Soc. Victoria 17: 74. 1881. Muell.-Arg. (1895b). Q.

*P. CONSPURCATA Muell.-Arg., Bull. Herb. Boiss. 3: 326. 1895. NSW.

*P. DEFOSSA Muell.-Arg., Flora 65: 518. 1882. Muell.-Arg. (1895b); Shirley (1890). Q.

*P. FIBRATA (Stirt. ex Bailey) A. Zahlbr., Cat. Lich. Univ. 1: 430. 1922. Verrucaria fibrata Stirt. ex Bailey, Queensland Agr. J. 5: 488. 1899. Q.

*P. FINITIMA Muell.-Arg., Flora 70: 429. 1887. Muell.-Arg. (1895b); Shirley (1889b, 1890). Q.

*P. IMMERSA Muell.-Arg., ibid. Muell.-Arg. (1895b); Shirley (1889b, 1890). Q.

*P. INDUSIATA Muell.-Arg., Rep. Austral. Assoc. Adv. Sci., p. 456. 1895. Shirley (1895). Q.

P. KUNTHII (Fée) Fée, Suppl. Essai Crypt. Ecorc. Officin. p. 80. 1837. Verrucaria kunthii Fée, loc. cit. p. 88. 1824. Muell.-Arg. (1893, 1895b); Shirley (1893a). Q, V.

P. MAMILLANA (Ach.) Trev., Conspect. Verruc. p. 13. 1860. Verrucaria mamillana Ach., Method. Lich. p. 120. 1803. Bailey (1891a); Muell.-Arg. (1891b, 1892a, 1893, 1895b); Shirley (1891a). Q, V.

P. MARGINATA Hook. in Kunth, Syn. Pl. Aequin. Orb. Nov. 1: 20. 1822. Muell.-Arg. (1895b); Shirley (1893a). Q.

P. MASTOPHORA (Nyl.) Muell.-Arg., Flora 66: 246. 1883. Verrucaria mastophora Nyl., Ann. Sci. Nat., Bot. (4) 15: 52. 1861. *V. flaventior Stirt., Trans. Proc. Roy. Soc. Victoria 17: 74. 1891 Q. Pyrenula flaventior Muell.-Arg., Flora 70: 429. 1887. Bailey (1881, 1883); F. Mueller (1881); Muell-Arg. (1891a, 1895b); Shirley (1890, 1891a). Q.

P. MASTOPHORIZANS Muell.-Arg., op. cit. 68: 515. 1885. Shirley (1895). Q.

*P. MELALEUCA Muell.-Arg., Nuovo Giorn. Bot. Ital. 30: 403. 1891. Bailey (1891); Muell.-Arg. (1895b); Shirley (1891b). Q.

*P. MICROCARPOIDES Muell.-Arg., Rep. Austral. Assoc. Adv. Sci. p. 457. 1895. Shirley (1895). Q.

*P. NIGROCINCTA Muell.-Arg., Nuovo Giorn. Bot. Ital. 23.: 404. 1891. Bailey (1891); Muell.-Arg. (1895b); Shirley (1891b). Q.

P. NITIDA (Weig.) Ach., Syn. Lich. p. 125. 1814. Sphaeria nitida Weig., Observ. Bot. p. 45. 1772. Muell.-Arg. (1893, 1895b); Shirley (1893a). Q, V.

*P. NITIDANS Muell.-Arg., Hedwigia 30: 56. 1891. Bailey (1891a); Muell.-Arg. (1895b); Shirley (1891b). Q.

P. OCCULTA (Knight) Muell.-Arg., Bull. Soc. Bot. Belg. 31 (2) 41. 1892. Verrucaria occulta Knight, Trans. Linn. Soc. London, Bot. 1: 279. 1877. Pyrenula micromma Shirley, Lich. Fl. Queensland. 4: 180. 1890 (non Trevis.). Q.

*P. OXYSPORIZA (Muell.-Arg.) A. Zahlbr., Cat. Lich. Univ. 1: 449. 1922. P. oxyspora Muell.-Arg., Rep. Austral. Assoc. Adv. Sci. p. 456. 1895. (non Hepp). Shirley (1895). Q.

P. PINGUIS Fée, Essai Crypt. Ecorc. Officin. p. 75. 1824. Muell-Arg. (1895b); Shirley (1893a). Q.

P. PINGUIS var. EMERGENS (Muell.-Arg.) Muell.-Arg., Bull. Soc. Bot. Belg. 30: 95. 1891. P. punctella var. emergens Muell.-Arg., Rév. Mycol. 10: 184. 1888. Muell.-Arg. (1895b); Shirley (1895). Q.

P. PORINOIDES Ach., Syn. Lich. p. 128. 1814. Muell.-Arg. (1893, 1895b); Shirley (1895). Q.

*P. SEGREGATA (Nyl.) Muell.-Arg., Flora 70: 429. 1887. Verrucaria aggregata f. segregata Nyl., Bull. Soc. Linn. Normand. (2) 2: 128. 1868. Muell.-Arg. (1895b); Shirley (1889b, 1890). Q.

P. SEXLOCULARIS (Nyl.) Muell.-Arg., Flora 60: 475. 1877. Verrucaria sexlocularis Nyl., Ann. Sci. Nat., Bot. (4) 20: 249. 1863. Muell.-Arg. (1895b); Shirley (1895). Q.

*P. SUBCONGRUENS Muell.-Arg., Rep. Austral. Assoc. Adv. Sci. p. 457. 1895. Shirley (1895). Q.

P. VELATIOR Muell.-Arg., Flora 68: 334. 1885. Muell.-Arg. (1895b); Shirley (1895). Q.

P. WARMINGII (Kremp.) Muell.-Arg., op. cit. 67.: 664. 1884. Verrucaria warmingii Kremp., Vidensk. Meddel. Dansk Naturhist. Foren. Kjoebenhavn 5: 395. 1874. Muell.-Arg. (1891a, 1895b); Shirley (1891a). Q.

PYRGILLUS

*P. AUSTRALIENSIS F. Wils. ex Bailey, Queensland Dept. Agr. Bull. 7: 31. 1891. Q.

*P. CALICIISPORUS F. Wils. ex Bailey, ibid. Q.

*P. FALLAX F. Wils. ex Bailey, loc. cit., p. 32. 1891. Q.

P. JAVANICUS (Mont. & v. d. Bosch) Nyl., Mém. Soc. Scienc. Nat. Cherbourg 5: 334. 1847. Calicium javanicum Mont. & v. d. Bosch in Jungh., Pl. Jungh., fasc. 4: 480. 1855. Muell.-Arg. (1891a); Shirley (1891a, 1893a). Q.

PYXINE

P. COCOES (Sw.) Nyl., Mem. Soc. Imp. Scienc. Nat. Cherbourg 5: 108. 1857. Lichen cocoes Sw., Nov. Gen. Sp. Pl., p. 146. 1788. Bailey (1883); Krempelhuber (1868); F. Mueller (1881). Q.

P. COCOES var. ENDOXANTHA Muell.-Arg., Flora 65: 318. 1882. Bailey (1891a); Muell.-Arg. (1891b); Shirley (1891a). Q.

"P. CONFLUENS Fr." Bailey (1883); Stirton (1899b). Q. We are unable to find this epithet, possibly Physcia confluens (=P. aegialita).

P. ENDOCHRYSINA Nyl., Lich. Japon. p. 34. 1890. Shirley (1893a). Q.

P. ENDOLEUCA (Muell.-Arg.) Vain., Hedwigia 37: (42). 1898. P. meissneri var. endoleuca Muell.-Arg., Flora 62: 290. 1879. Shirley (1893a). Q.

P. ESCHWEILERI (Tuck.) Vain., Etud. Lich. Brésil 1: 56. 1890. P. cocoes var. eschweileri Tuck., Proc. Amer. Acad. Arts Sci. 12: 167. 1877. P. sorediata Tuck., Syn. N. Am. Lich. 1: 80. 1882. Bailey (1881); Shirley (1889a). Q.

P. MEISSNERI Tuck. ex Nyl., Ann. Sci. Nat., Bot. (4) 11: 205. 1859. Bailey (1881, 1883); F. Mueller (1881); Shirley (1889a); Stirton (1899b). Q.

P. MEISSNERI var. SOREDIOSA Muell.-Arg., Flora 62: 290. 1879. Shirley (1893a). Q.

*P. OBSCURIOR Stirt., Trans. Proc. Roy. Soc. Victoria 17: 70. 1881. Bailey (1881, 1883); F. Mueller (1881); Shirley (1889a); Stirton (1899b). Q.

P. RETIRUGELLA Nyl., Ann. Sci. Nat., Bot. (4) 11: 240. 1859. Bailey (1891a); Muell.-Arg. (1891b); Shirley (1889a). Q.

*P. RETIRUGELLA var. ENDOXANTHA Muell.-Arg., Bull. Herb. Boiss. 4: 91. 1896. Q.

P. RETIRUGELLA f. SOREDIOSA Muell.-Arg., ibid. Q.

*P. RUGULOSA Stirt., Trans. Proc. New Zealand Inst. 30: 396 (1897) 1898. Stirton (1899b). Q.

*P. SUBCINEREA Stirt., loc. cit. p. 397. Stirton (1899b). Q.

*P. SUBVELATA Stirt., loc. cit. p. 396. Stirton (1899b). Q.

RAMALEA

*R. COCHLEATA Muell.-Arg., Bull. Herb. Boiss. 4: 88. 1896. V. *Thysanothecium hyalinum f. squamosulum F. Wils., Proc. Roy. Soc. Victoria (2) 5: 176. 1892 (non "squamosum" as given by Zahlbruckner, Cat. Lich. Univ. 6: 606. 1930). Muell.-Arg. (1898). Q, V.

RAMALINA

R. ANCEPS Nyl., Syn. Lich. 1: 291. 1860. Nylander (1857, as "Alectoria anceps Nyl.," nomen nudum?). Sine loc.

R. ANGULOSA Laur. ex Th. Fr., Flora 44: 411. 1861. NSW.

*R. AUSTRALIENSIS Nyl., Bull. Soc. Linn. Normand. (2) 4: 120. 1870. WA.

*R. BREVIS F. Wils., Victoria Naturalist 6: 69. 1889 (overlooked by Zahlbruckner, Cat. Lich. Univ.). V.

*R. CALICARIS (L.) Roehl., Deutschl. Fl. 3 (2): 139. 1813. Lichen calicaris L., Sp. Pl., p. 1146. 1753. Bailey (1883); Bibby (1959); Bibby & Smith (1954); Knight (1882); Laurer (1827); F. Mueller (1881); Shirley (1889a); Tate (1887); Watts (1903); Willis (1953); Wilson (1889c). NSW, Q, SA, V, WA.

*R. CALICARIS var. AUSTRALICA Raes., Arch. Soc. Zool. Bot. Fenn. "Vanamo" 3: 178. 1949. V.

R. COMPLANATA (Sw.) Ach., Lichenogr. Univ. p. 599. 1810. Lichen complanatus Sw. ex Ach., Kongl. Svenska Vetensk.-Akad. Nya Handl. p. 290. 1797. Bailey (1881, 1883); Hue (1890-92); Nylander (1870); F. Mueller (1881); Shirley (1889a); Zahlbruckner (1896). NSW, Q.

*R. CONFIRMATA Nyl., Bull. Soc. Linn. Normand. (2) 4: 138. 1870. Nylander (1870). WA.

R. DENDRISCOIDES Nyl., Flora 59: 412. 1876. Muell.-Arg. (1891a); Shirley (1891a). Q.

R. DENDRISCOIDES var. MINOR Muell.-Arg., Proc. Roy. Soc. Edinburgh 11: 458. 1882. Muell.-Arg. (1891a); Shirley (1891a). Q.

*R. DILACERATA (Hoffm.) Vain. var. ALBA A. Zahlbr., Cat. Lich. Univ. 6: 462. 1930. R. minuscula var. alba Knight ex Shirley, Proc. Roy. Soc. Queensland 5: 103. 1888. Shirley (1892b). Q.

R. ECKLONII (Spreng.) Mey. & Flot., Nova Actorum Acad. Caes. Leop.-Carol. Nat. Cur. 19, Suppl.: 213. 1843. Parmelia ecklonii Spreng. Syst. Veg. 4(2): 328. 1827. Ramalina yemensis Nyl., Bull. Soc. Linn. Normand. (2) 4: 144. 1870. Bailey (1883); Hellbom (1896); Hue (1890-92) Krempelhuber (1880); F. Mueller (1881); Muell.-Arg. (1891a); Nylander (1857, 1870, 1886); Shirley (1889a, 1892b); Tate (1887); Willis (1953). NSW, Q, SA.

*R. ECKLONII var. MEMBRANACEA (Laur.) Muell.-Arg., Bull Herb. Boiss. 2, Appendix I: 30. 1894. *R. fraxinea var. membranacea Laur., Linnaea 2: 43. 1827. Hampe (1853); Muell.-Arg. (1891a, 1892a); Shirley (1891a). Q.

R. ECKLONII var. OVALIS F. Wils., Pap. Proc. Roy. Soc. Tasmania for 1892: 159. 1893. Stirton (1899b). Q.

R. ECKLONII var. TENUISSIMA Mey. & Flot., Nova Actorum Acad. Caes. Leop.-Carol. Nat. Cur. 19, Suppl.: 213. 1843. Shirley (1893a). Q.

*R. EXIGUELLA Stirt., Trans. Proc. Roy. Soc. Queensland 17: 68. 1881. Bailey (1881, 1883); F. Mueller (1881); Shirley (1889a). Q.

*R. FARINACEA (L.) Ach. var. DENDROIDES Muell.-Arg., Flora 66: 21. 1883. Muell.-Arg. (1892a). Q.

*R. FARINACEA var. NERVULOSA Muell.-Arg., op. cit. 60: 21. 1883 (not nervosula as given by Zahlbruckner, Cat. Lich. Univ. 6: 471. 1930). Muell.-Arg. (1891a); Shirley (1889a). Q.

*R. FARINACEA var. SQUARROSA Muell.-Arg., ibid. Shirley (1890). Q.

R. FASTIGIATA (Pers.) Ach., Lichenogr. Univ. p. 603. 1810. Lichen fastigiatus Pers., Ann. Bot. (Usteri) 1: 156. 1794. Ramalina calicaris var. fastigiata Fr., Lichenogr. Europ. Reform. p. 30. 1831. Laurer (1827); Nylander (1857); Wilson (1887). V.

R. FRAXINEA (L.) Ach., Lichenogr. Univ. p. 602. 1810. Lichen fraxineus L., Sp. Pl. p. 1146. 1753. Ramalina calicaris var. fraxinea Mont. in Webb, Hist. Nat. Iles Canar. 3(2): 99. 1840. Nylander (1857); Shirley (1889a); Tate (1887); Watts (1903); Weber (1971); Wilson (1887, 1889c, 1890b). Q, SA, V, WA.

R. FRAXINEA var. AMPLIATA (Ach.) Ach., Lichenogr. Univ. p. 603. 1810. Parmelia fraxinea var. ampliata Ach., Method. Lich. p. 259. 1803. Wilson (1887). V.

R. FRAXINEA f. PLATYNA Nyl., Bull. Soc. Linn. Normand. (2) 4: 136. 1870. Shirley (1894, giving R. brevis F. Wils. as synonym). V.

E. FRAXINEA var. TAENIATA (Ach.) Rebent., Prodrom. Fl. Neomarch. p. 309. 1804. Parmelia fraxinea var. taeniata Ach., Method. Lich.

p. 259. 1803. Ramalina fraxinea var. taeniaeformis Ach., Lichenogr. Univ. p. 603. 1810. Wilson (1887). V.

R. FURCELLATA A. Zahlbr., Cat. Lich. Univ. 6: 488. 1930. Evernia furcellata Mont. in Sagra, Hist. l'Ile Cuba, Bot. p. 236. 1838-42. Ramalina gracilenta Fries, Lichenogr. Europ. Reform. p. 29. 1831 (non Roehl., 1813). F. Mueller (1881). NSW, V.

R. GENICULATA Hook. f. & Tayl., London J. Bot. 3: 655. 1844. Crombie (1880); Hellbom (1896); Hue (1890-92); Nylander (1870); Shirley (1889a); Tate (1887). NSW, Q, SA.

*R. GENICULATA var. COMPACTA Muell.-Arg., Bull. Herb. Boiss. 4: 88. 1896. Muell.-Arg. (1898); Weber (1971). Q, V.

R. GENICULATA var. OLIVACEA Muell.-Arg., Flora 62: 294. 1879. Shirley (1893a, as R. "inflata" var. olivacea Muell.-Arg. and citing Lich. Beitr. No. 128, which is the above).

*R. GLAUCESCENS Kremp., Verh. zool.-bot. Ges. Wien 30: 333. 1880. F. Mueller (1880). V.

R. GRACILIS (Pers.) Nyl., Syn. Lich. 1: 296. 1860. Physcia gracilis Pers. in Gaudich., Voy. Uranie, Bot. p. 209. 1826. Bailey (1891a); Hue (1890-92); Muell.-Arg. (1891b); Nylander (1870); Shirley (1889a); Turner (1905). NSW, Q.

R. HOMALEA Ach., Lichenogr. Univ. p. 598. 1810. Nylander (1857). Sine loc., report in doubt. R. homalea is a North American endemic.

R. INFLATA Hook. f. & Tayl. in Hook., Fl. Antarct. 1: 194. 1844. Krempelhuber (1880); F. Mueller (1881); Muell.-Arg. (1887c, 1891a, 1892a); Shirley (1889b); Stirton (1899b). NSW, Q, V.

R. INFLATA var. GRACILIS (Bab.) Muell.-Arg., Flora 71: 131. 1888. R. calicaris var. gracilis Bab. ex Hook., Fl. Nov. Zel. 2: 270. 1855. Shirley (1889b, 1894). Q.

R. INTERMEDIA Del. ex Nyl., Bull. Soc. Linn. Normand. (2) 4: 166. 1870. Shirley (1888a, 1889a). Q.

R. JAVANICA Nyl., loc. cit., p. 167. F. Mueller (1881). Muell.-Arg. (1887c, relegating F. Mueller's report to R. farinacea var. dendroides Muell.-Arg.). NSW.

*R. KNIGHTIANA A. Zahlbr., Cat. Lich. Univ. 6: 494. 1830. NSW. R. subgeniculata Knight, Trans. Linn. Soc. London, Bot. 2: 50. 1882 (non Nyl., 1870). Wilson (1889a). NSW, V.

*R. LACERATA Muell.-Arg., Flora 66: 20. 1883. WA.

*R. LEIODEA Nyl., Bull. Soc. Linn. Normand. (2) 4: 141. 1870. Bailey (1891a); Hue (1890-92); Muell.-Arg. (1891b). Shirley (1891a, 1892b). Q.

*R. LEIODEA Nyl. var. FASTIGIATULA Muell.-Arg., Flora 66: 21. 1883. V. Bibby (1959), NT, V.

R. LINEARIS (Sw.) Ach., Lich. Univ. p. 598. 1810. Lichen linearis Sw., Meth. Musc. p. 36. 1781. Darbishire (1912). Sine loc.

*R. MYRIOCLADA Muell.-Arg., Flora 63: 20. 1883. NSW.

*R. PERPUSILLA Stirt., Trans. Proc. Roy. Soc. Victoria 17: 68. 1881. Bailey (1881, 1883); F. Mueller (1881); Shirley (1889a). Q.

R. POLLINARIA (Liljebl.) Ach., Lichenogr. Univ. p. 608. 1810. Lichen calicaris C. pollinarius Liljebl., Utkast Svensk Fl. p. 426. 1792. Shirley (1894). V.

R. POLYMORPHA (Liljebl.) Ach., Lichenogr. Univ. p. 600. 1810. Lichen calicaris var. polymorphus Liljebl., Utkast Svensk Fl. p. 426. 1792. Hue (1890-92). Sine loc.

R. POLYMORPHA f. EMPLECTA (Ach.) Ach., Lichenogr. Univ. p. 601. 1810. Parmelia polymorpha var. emplecta Ach., Method. Lich. p. 267. 1803. Shirley (1894). V.

R. PUSILLA Le Prev. in Duby, Bot. Gall. 2: 614. 1830. Fries (1846-47); Krempelhuber (1880); F. Mueller (1881); Muell.-Arg. (1887c); Nylander (1857); Shirley (1889a); Tate (1881); Q, SA, V.

"R. RUTILANS Stirt." Shirley (1889a, error for Stictina rutilans); Bailey (1883).

R. SCOPULORUM (Retz.) Ach., Lichenogr. Univ. p. 604. 1810. Lichen scopulorum Retz., Fasc. Obs. Bot. 4: 30. 1786. Bailey (1883); Krempelhuber (1880); F. Mueller (1881); Muell.-Arg. (1887c); Nylander (1857); Shirley (1889a); Wilson (1889b). NSW, Q.

R. SCOPULORUM var. CUSPIDATA Ach., Lichenogr. Univ. p. 605. 1810. Bailey (1881). Q. Equals R. siliquosa (Huds.) A.L. Sm. fide Zahlbruckner.

R. SCOPULORUM var. SUBFARINACEUM F. Wils., Proc. Roy. Soc. Queensland 6: 90. 1889. Nomen nudum.

R. TAYLORIANA A. Zahlbr., Cat. Lich. Univ. 6: 524. 1930. Ramalina canaliculata Tayl., London J. Bot. 6: 188. 1847 (non Dietr., 1812). Wilson (1889a). V.

*R. UNILATERALIS F. Wils., Victoria Naturalist 6: 69. 1889 (overlooked by Zahlbr., Cat. Lich. Univ.). V. = R. pollinaria, fide Shirley (1894).

R. USNEA (L.) Howe f., Bryologist 17: 81. 1914. Lichen usnea L., Mant. 1: 131. 1767. Ramalina usneoides Mont. in Gaudich., Voyage Bonité p. 156. 1844-46. Hellbom (1896); Hue (1890-92); Nylander (1870); Wilson (1889b). NSW.

RAMALODIUM

*R. SUCCULENTUM Nyl. ex Crombie, J. Linn. Soc. London, Bot. 17: 392. 1879. Henssen (1965); Hue (1890-92). NSW.

RHIZOCARPON

R. BADIOATRUM (Flk.) Th. Fr. var. ATROALBUM (L.) Malme ex Vain., Acta Soc. Faun. Fl. Fenn. 53: 328. 1922. Lichen atroalbus L., Sp. Pl. p. 1141. 1753. F. Mueller (1881); Muell.-Arg. (1893); Shirley (1889a). Q, V.

R. BADIOATRUM var. INCUSUM (Ach.) A. Zahlbr., Cat. Lich. Univ. 4: 327. 1927. Lecidea incusa Ach., Synops. Lich., p. 33. 1814. *Rhizocarpon rivulare Muell.-Arg., Bull. Herb. Boiss. 1: 53. 1893. V.

*R. CLAUSUM (Knight ex Shirley) A. Zahlbr., Cat. Lich. Univ. 4: 350. 1927. Lecidea clausa Knight ex Shirley. Proc. Roy. Soc. Queensland 6: 183. 1889. Shirley (1888a). Q.

R. GEOGRAPHICUM (L.) DC. in Lam. & DC., Flor. Franç. ed. 3, 2: 365. 1805. Lichen geographicus L., Sp. Pl. p. 1140. 1753. Hellbom (1896); Hue (1890-92); Nylander (1886); Willis (1953); Wilson (1887, 1890a). NSW, V, WA.

R. GEOGRAPHICUM (L.) DC. f. CYCLOPICUM Flagey, Mém. Soc. d'Emulat. Doubs (6) 8: 91. 1894. Lecidea geographica var. cyclopica Nyl., Lich. Scand. p. 248. 1861. Shirley (1889a,e). Q.

R. PETRAEUM (Wulf.) Mass., Ricerch. Auton. Lich. p. 102. 1852. Lichen petraeus Wulf., Schrift. Ges. Naturf. Fr. Berlin 8: 89. 1787. Hue (1890-92); Nylander (1857). Sine loc.

R. SUPERFICIALE (Schaer.) Vain. ssp. splendidum (Malme) Runem., Opera Bot. 2(1): 55. 1956. R. splendidum Malme, Sv. Bot. Tidskr. 20: 56. 1926. R. crystalligenum Lynge, Skr. Svalb. Ishavet 47: 19. 1932. NSW: Kosciusko State Park, trail to Northcote Pass, 7000 ft., 10 Nov. 1967, Weber & McVean L-47178. (COLO).

RINODINA

R. AEQUATA (Ach.) Flagey, Cat. Lich. Algérie, p. 40. 1896. Lecanora coniops var. aequata Ach., Lichenogr. Univers. p. 171. 1810. Hue (1890-92).

*R. AUSTRALIENSIS Muell.-Arg., Hedwigia 32: 123. 1893. Shirley (1894). Q, V.

R. BISCHOFFII (Hepp) Mass., Framm. Lich. p. 26. 1855. Psora bischoffii Hepp, Flecht. Europ. no. 81. 1853. Muell.-Arg. (1892b). WA.

R. COLOBINOIDES (Nyl.) A. Zahlbr., Cat. Lich. Univ. 7: 499. 1931. Lecanora colobinoides Nyl., Acta Soc. Sci. Fenn. 7: 444. 1863. Muell.-Arg. (1893). V.

*R. DIFFRACTELLA Muell.-Arg., Bull. Herb. Boiss. 3: 634. 1895. Shirley (1896). Q.

R. EXIGUA (Ach.) S. Gray, Nat. Arr. Brit. Pl. 1: 450. 1821. Lichen exiguus Ach., Lichenogr. Suec. Prodrom., p. 69. 1798. Lecanora exigua Roehl., Deutschl. Fl. 3 (2): 72. 1813. Shirley (1889a). Q.

R. GLOMERELLA (Stirt.) A. Zahlbr., Cat. Lich. Univ. 7: 520. 1931. The basionym is Lecidea glomerella Stirt., not Lecanora glomerella Stirt., as given by Zahlbruckner. The plant is obviously a Buellia from the description and discussion.

R. METABOLICA (Ach.) Anzi, Comment. Soc. Crittogam. Ital. 2(1): 10. 1864. Lecanora metabolica Ach., Lichenogr. Univ. p. 351. 1810. Muell.-Arg. (1893). V.

R. METABOLICA var. PHAEOCARPA Muell.-Arg., Rév. Mycol. 10: 63. 1888. Muell.-Arg. (1893). V.

R. MINUTULA Muell.-Arg., Flora 62: 291. 1879. Muell.-Arg. (1891a), Shirley (1891a). Q.

*R. OBSCURA Muell.-Arg., Bull. Herb. Boiss. 1: 40. 1893. V.

*R. PACHYSPORA Muell.-Arg., ibid. V.

R. PLACOMORPHA (Stirt.) A. Zahlbr., Cat. Lich. Univ. 7: 543. 1931. The basionym is Lecidea placomorpha Stirton ex Bailey, Queensland Agr. J. 5: 487. 1899, not Lecanora as given by Zahlbruckner, and the plant is obviously a Buellia from the description and the discussion of the new species in connection with B. disciformis.

*R. SUBCRUSTACEA (Muell.-Arg.) A. Zahlbr., Cat. Lich. Univ. 7: 567. 1931. Physcia subcrustacea Muell.-Arg., Bull. Herb. Boiss. 1: 33. 1893. V.

R. THIOMELA (Nyl.) Muell.-Arg., Flora 64: 515. 1881. Lecanora thiomela Nyl., op. cit. 48: 338. 1865. Muell.-Arg. (1893). Shirley (1889a). Q, V.

*R. XANTHOMELANA Muell.-Arg., Nuovo Giorn. Bot. Ital. 23: 390. 1891. Shirley (1891b). Q.

ROCCELLA

R. MONTAGNEI Bel., Voy. Indes-Orient. 2: 117. 1846. Crombie (1880); Bibby (1959); Darbishire (1898). NT. Reports of R. tinctoria (New Zealand) and R. phycopsis (New Caledonia) were given by Darbishire and erroneously attributed to Australia by Zahlbruckner in Engl.-Prantl, Nat. Pflanzenfam., ed. 2, 8: 126. 1926.

SAGIOLECHIA

*S. LEPTOPLACELLA (Muell.-Arg.) A. Zahlbr., Cat. Lich. Univ 2: 733. 1924. Patellaria leptoplacella Muell.-Arg., Bull. Herb. Boiss. 4: 95. 1896. Muell.-Arg. (1898). Q.

SARCOGRAPHA

*S. COLLICULOSA (Knight ex Bailey) A. Zahlbr., Cat. Lich. Univ. 2: 459. 1923. Glyphis colliculosa Knight ex Bailey, Syn. Queensland Fl. 1: 75. 1886. Shirley (1889a). Q.

*S. KIRTONIANA (Muell.-Arg.) Muell.-Arg., Flora 70: 77. 1887. Glyphis (non Graphis as given by Zahlbruckner, Cat. 2: 462. 1923) kirtoniana Muell.-Arg., op. cit. 63: 516. 1882. NSW. Wilson (1888). NSW, V.

S. LABYRINTHICA (Ach.) Muell.-Arg., Mém. Soc. Phys. Hist. Nat. Genève 29(8): 62. 1887. Glyphis labyrinthica Ach., Syn. Lich. p. 107. 1814. Bailey (1881, 1883, 1891a); F. Mueller (1881); Muell-Arg. (1891b, 1893); Shirley (1889a). Q.

S. MEDUSULA (Spreng.) Fée, Nova Actorum Acad. Caes. Leop.-Carol. Nat. Cur. 18, suppl. 1, p. 22. 1841. Asterisca medusula Spreng., Syst. Veg. 4 (1): 254. 1827. Krempelhuber (1880); F. Mueller (1881). Q.

S. MEDUSULINA (Nyl.) Muell.-Arg., Flora 70: 77. 1887. Glyphis medusulina Nyl., Acta Soc. Scient. Fenn. 7: 485. 1863. Bailey (1881, 1883); F. Mueller (1881); Shirley (1889a). Q.

*S. OCULATA Muell.-Arg., Bull. Herb. Boiss. 3: 323. 1895. Shirley (1896). Q.

*S. SUBTRICOSA Muell.-Arg., Flora 70: 78. 1887. Q. *Sarcographa actinota F. Wils. ex Bailey, Queensland Dept. Agr. Bull. 7: 33. 1891. Q. Shirley (1893a).

*S. SUBTRICOSA var. PULVERULENTA (F. Wils.) A. Zahlbr., Cat. Lich. Univ. 2: 466. 1923. *Sarcographa actinota var. pulverulenta F. Wils. ex Bailey, Queensland Dept. Agr. Bull. 7: 33. 1891. Q.

SARCOGRAPHINA

*S. CYCLOSPORA (Shirley) Muell.-Arg., Flora 70: 425. 1887. Glyphis cyclospora Shirley, Proc. Roy. Soc. Queensland 6: 215. 1889. Shirley (1889a,b). Q.

SARCOGYNE

S. REGULARIS Koerb., Syst. Lich. German. p. 267. 1855. S. pruinosa Koerb. var. minuta Mass., Sched. Crit. 10: 177. 1856. Muell.-Arg. (1892b). WA. For nomenclature of S. pruinosa and regularis see P. James, Lichenologist 3: 145. 1965.

(LECIDEA) SIMPLEX var. CALCIFRAGA Muell.-Arg., Rév. Mycol. 2: 79. 1880. Muell.-Arg. (1893). V.

SCHISMATOMMA

*S. (?) ALBOVESTITUM (Knight) A. Zahlbr., Cat. Lich. Univ. 2: 553. 1923. Platygrapha (?) albovestita Knight, Trans. Linn. Soc. London, Bot. 2: 43. 1882. NSW. Wilson (1889a,b,c). NSW, V.

*S. BANKSIAE (Muell.-Arg.) A. Zahlbr., Cat. Lich. Univ. 2: 554. 1923. Platygrapha banksiae Muell.-Arg., Bull. Herb. Boiss. 1: 55. 1893. V.

*S. DIRINEUM (Nyl.) Raes. var. AUSTRALICA Raes., Ann. Soc. Zool.-Bot. Fenn. "Vanamo" 21, 16: 6. 1946. V.

*S. NANOCARPUM (Knight) A. Zahlbr., Cat. Lich. Univ. 2: 560. 1923. Stigmatidium nanocarpum Knight, Trans. Linn. Soc. London, Bot. 2: 42. 1882. Platygrapha leptospora Muell.-Arg., Bull. Herb. Boiss. 3: 316. 1895. NSW.

S. OCELLATUM (Nyl.) A. Zahlbr., Cat. Lich. Univ. 2: 560. 1923. Platygrapha ocellata Nyl., Acta Soc. Scient. Fenn. 7: 478. 1863. Nylander (1886). NSW.

*S. SHIRLEYANUM (Muell.-Arg.) A. Zahlbr., Cat. Lich. Univ. 2: 564. 1923. Platygrapha shirleyana Muell.-Arg., Bull. Herb. Boiss. 3: 316. 1895. Shirley (1896). Q.

SIPHULA

*S. CORIACEA Tayl. ex Nyl., Syn. Lich. 1: 263. 1860. WA. *S. caesia Muell.-Arg., Hedwigia 31: 191. 1892. WA. Bibby & Smith (1954); Frey (1967); Hue (1890-92); Kurokawa (1971); Weber (1969). NSW, WA.

SPHAEROPHORUS

"S. CORALLOIDES Pers." Reports are misidentifications of Stereocaulon ramulosum, fide Muell.-Arg. (1887c).

S. MELANOCARPUS (Sw.) DC. in Lam. & DC., Flor. Franç., ed. 2, 6: 178. 1805. Lichen melanocarpus Sw., Nov. Gen. Sp. Pl. p. 147. 1788. Sphaerophoron compressum Ach., Method. Lich. p. 135. 1803. Bailey (1883, 1891a); Bibby (1954); Hellbom (1896); Hue (1890-92); F. Mueller (1881); Muell.-Arg. (1881b, 1891b); Nylander (1857); Shirley (1889a); Turner (1905); Wilson (1892); Zahlbruckner (1896). NSW, Q. V.

S. MELANOCARPUS var. AUSTRALIS (Laur.) Murray, Trans. Roy. Soc. New Zealand 88: 188. 1960. *Sphaerophoron australe Laur., Linnaea 2: 44. 1827. Hellbom (1896); Hue (1890-92); Muell.-Arg. (1891a, 1892a); Shirley (1891a); Wilson (1887, 1891b, 1892); Zahlbruckner (1896). NSW, Q, V.

S. MELANOCARPUS var. AUSTRALIS f. ANGUSTIOR (Reinke) Murray, Trans. Roy. Soc. New Zealand 88: 190. 1960. S. australe f. angustior Reinke, Jahrb. Wiss. Bot. 28: 85. 1895. Murray (1960). Sine loc.

S. MELANOCARPUS var. AUSTRALIS f. SUBTERES (A. Zahlbr.) Murray, Trans. Roy. Soc. New Zealand 88: 191. 1960. S. australis f. subteres A. Zahlbr. ex H. Magn., Ark. Bot. 31A (1): 24. 1944. Murray (1960). Sine loc.

(Sphaerophorus)

*S. MELANOCARPUS var. CANDIDUS (Muell.-Arg.) A. Zahlbr., Cat. Lich. Univ. 1: 695. 1922. Sphaerophoron compressum var. candidum Muell.-Arg., Flora 64: 505. 1881. Q. Muell.-Arg. (1892a); Shirley (1889b, 1890). Q.

*S. MELANOCARPUS f. INSIGNIS (Laur.) Murray, Trans. Roy. Soc. New Zealand 88: 190. 1960. S. australis f. insignis Muell.-Arg., Flora 66: 17. 1883. Sphaerophoron insigne Laur., Linnaea 2: 45. 1827. *S. ceranoides Hampe, Linnaea 28: 217. 1856. V.

*S. MELANOCARPUS f. PROLIFERUS (F. Wils.) Murray, Trans. Roy. Soc. New Zealand 88: 191. 1960. S. australis var. proliferus F. Wils., J. Linn. Soc. London, Bot. 28: 370. 1891. Wilson (1892). V.

S. TENER Laur., Linnaea 2: 45. 1827 (as Sphaerophoron tenerum). Baker (1902); Darbishire (1912); Krempelhuber (1870, 1880); F. Mueller (1881); Muell.-Arg. (1887c, 1892); Murray (1960a). NSW, V.

SPHINCTRINA

S. MICROCEPHALA (Sm.) Nyl., Mém. Imp. Soc. Scienc. Nat. Cherbourg 3: 168. 1855. Lichen microcephalus Sm. in Sm. & Sowerb., Engl. Bot. 26: tab. 1865. 1808. Wilson (1889b, 1892). NSW, V.

*S. MICROCEPHALA var. TENELLA F. Wils., Victoria Naturalist 6: 63. 1889, J. Linn. Soc. London, Bot. 28: 361. 1891. Wilson (1889a, 1892). V.

SPORASTATIA

S. TESTUDINEA (Ach.) Mass., Geneac. Lich. p. 9. 1854. Lecidea cechumena var. testudinea Ach., Kongl. Svenska Vetensk.-Akad. Nya Handl. p. 232. 1808. NSW: Mount Kosciusko State Park, trail to Northcote Pass, 7000 ft., 10 Nov. 1967, Weber & McVean L-47190 pr. p. (COLO).

SPOROPODIUM

S. PHYLLOCHARIS (Mont.) Mass. var. FLAVESCENS R. Sant., Symb. Bot. Upsal. 12(1): 518. 1952. Biatora phyllocharis Mont., Lecidea phyllocharis Nyl., of Australian reports. Bailey (1881, 1883); F. Mueller (1881); Muell.-Arg. (1890b); Santesson (1952); Shirley (1888b, 1889a). NSW, Q.

S. XANTHOLEUCUM (Muell.-Arg.) A. Zahlbr., Cat. Lich. Univ. 2: 681. 1924. Gyalectidium xantholeucum Muell.-Arg., Flora 64: 101. 1881. Santesson (1952). Q.

STEREOCAULON

S. CAESPITOSUM Redgr., Hedwigia 76: 132, 138. 1936. NSW: Mount Stilwell above Charlottes Pass, 6500 ft., 2 Feb. 1968, Weber & McVean L-49829,! I. M. Lamb (COLO).

S. CORTICATULUM Nyl., Flora 41: 117. 1858. Wilson (1889a). V.

*S. CORTICATULUM var. HUMILE (Muell.-Arg.) M. Lamb ex Frey, Bot. Jahrb. 86: 244. 1967. *S. humile Muell.-Arg., Bull. Herb. Boiss. 4: 88. 1896. V. Frey (1967); Muell.-Arg. (1898). V.

S. EXALBIDUM Nyl., Ann. Sci. Nat., Bot. (4) 11: 210. 1859. F. Mueller (1881). NSW, V.

S. LEPTALEUM Nyl., Syn. Lich. 1: 251. 1860. Wilson (1889a). V.

S. MIXTUM Nyl., Ann. Sci. Nat., Bot. (4) 11: 210. 1859. Muell.-Arg. (1892a); Zahlbruckner (1896). NSW.

(Stereocaulon)

S. PROXIMUM Nyl., ibid. F. Mueller (1881); Muell.-Arg. (1887c); Wilson (1890a). V.

S. RAMULOSUM (Sw.) Raeuschel, Nomencl. Bot., ed. 3, p. 328. 1797. Lichen ramulosus Sw., Nova Gen. Sp. Pl. p. 147. 1788. Bailey (1883); Crombie (1880); Krempelhuber (1880); F. Mueller (1881); Muell.-Arg. (1887c); Nylander (1857); Shirley (1889a); Turner (1905); Zahlbruckner (1896). NSW, Q, V.

*S. RAMULOSUM var. COMPACTUM Muell.-Arg., Bull. Herb. Boiss. 4: 87. 1896. Muell.-Arg. (1896). Q.

*S. RAMULOSUM var. MACROCARPOIDES Muell.-Arg., op. cit. 2, Appendix 1: 22. 1894. S. macrocarpoides Nyl., Syn. Lich. 1: 238. 1860. S. proximum var. macrocarpoides Nyl., Lich. Nov. Zel. p. 16. 1888. Shirley (1892a). Q.

*S. RAMULOSUM var. NUDATUM (Muell.-Arg.) Muell.-Arg., J. Linn. Soc. London, Bot. 32: 199. 1896. S. proximum var. nudatum Muell.-Arg., Flora 69: 252. 1886. "Brogers Creek." Bailey (1891a); Frey (1967); Muell.-Arg. (1891b); Shirley (1891a); Weber (1969). Q, V.

S. SUBRAMULOSUM Muell.-Arg., Flora 74: 108. 1891. Zahlbruckner (1896). NSW.

STICTA

*S. BREVIPES (Muell.-Arg.) A. Zahlbr., Cat. Lich. Univ. 3: 373. 1925. Stictina brevipes Muell. Arg., Flora 65: 302. 1882. Bailey (1883, 1891a); Cheel (1912-14); F. Mueller (1881); Muell-Arg. (1891a,b); Shirley (1893a). Q.

*S. CAMARAE Muell.-Arg. loc. cit. p. 303. NSW. Shirley (1889a). NSW, Q.

S. CELLULIFERA Hook. f. & Tayl., Flora Antarct. 1: 198. 1844. Stezenberger (1895). Sine loc.

*S. CELLULIFERA f. EXPALLIDA Stizenb., Flora 81: 114. 1895. Sticta fossulata f. expallida Kremp., Verh. zool.-bot. Ges. Wien 30: 336. 1880. V.

S. CINEREOGLAUCA Hook. f. & Tayl., London J. Bot. 3: 649. 1844. Hampe (1853); F. Mueller (1881). V.

S. CORIACEA Hook. f. & Tayl., loc. cit., p. 648. Ricasolia coriacea Nyl., Mém. Soc. Sci. Nat. Cherbourg 5: 103. 1857. Watts (1903). NSW.

*S. CYPHELLULATA (Muell.-Arg.) Hue, Nouv. Arch. Mus. Paris (4) 3: 99. 1901. Stictina cyphellulata Muell.-Arg., Flora 65: 301. 1882. Q. Bailey (1891a); Cheel (1912-14); Muell.-Arg. (1891b); Shirley (1889a); NSW, Q, V.

S. DAMAECORNIS (Sw.) Ach., Method. Lich. p. 276. 1803. Lichen damaecornis Sw., Nova Gen. Sp. Pl. p. 146. 1788. Hellbom (1896); Hue (1890-92); F. Mueller (1881); Nylander (1857); Wilson (1890a). V.

S. DEMUTABILIS Kremp., J. Mus. Godeffroy 1(4): 98. 1874. Bailey (1891a); Muell. Arg. (1891b); Shirley (1891a); Stizenberger (1895). Q.

S. DICHOTOMOIDES Nyl., Syn. Lich. 1: 355. 1860. Bailey (1891a); Cheel (1912-14); Muell.-Arg. (1891b); Shirley (1889a, 1894); Turner (1905). NSW, Q.

*S. DIVERSA (Stirt.) Stirt. ex A. Zahlbr., Cat. Lich. Univ. 3: 379. 1925. Stictina diversa Stirt., Trans. Proc. New Zealand Inst. 32: 75. 1899. Q.

S. DOZYANA Mont. & v. d. Bosch in Jungh., Pl. Jungh., Fasc. IV. p. 436. 1855. Shirley (1889a). Q.

S. FILICINA Ach., Method. Lich. p. 375. 1803 (pr. p.). Darbishire (1912); Shirley (1889a). Q.

S. FILIX (Sw.) Nyl., J. Linn. Soc. London, Bot. 9: 246. 1865. Lichen filix Sw., Method. Musc. p. 36. 1781. Sticta filicina Ach., Method. Lich. p. 275. 1803 (pr. p.). Bailey (1881, 1883); Cheel (1912-14); Krempelhuber (1880); F. Mueller (1881); Muell.-Arg. (1886); Turner (1905); Weber (1969); Wilson (1889b); Zahlbruckner (1896). NSW, Q.

*S. FILIX var. MYRIOLOBA Muell.-Arg., Flora 69: 254. 1886. Cheel (1912-14); Krempelhuber (1880, as f. minor, nomen); Muell.-Arg. (1887c, 1891a); Shirley (1892a). Q.

S. FULIGINOSA Ach., Method. Lich. p. 280. 1803. Cheel (1912-14); Shirley (1893a); Weber (1969); Wilson (1887). Q, V.

*S. GLAUCESCENS Kremp., Verh. zool.-bot. Ges. Wien 30: 334. 1880. *Sticta aurulenta Kremp., loc. cit. p. 335. NSW. F. Mueller (1881); Muell.-Arg. (1887c); Shirley (1891a); Stizenberger (1895). NSW, Q.

*S. IMPRESSULA (Nyl.) A. Zahlbr., Cat. Lich. Univ. 3: 388. 1925. Stictina tomentosa var. impressula Nyl., Ann. Sci. Nat., Bot. (5) 7: 305. 1867. S. impressula Nyl., Flora 57: 71. 1874. Shirley (1890); Stizenberger (1895). Sine loc.

*S. INSCULPTA (Stizenb.) A. Zahlbr., Cat. Lich. Univ. 3: 388. 1925. Stictina impressula Muell.-Arg., Flora 71: 22. 1888 (non Nyl.). S. insculpta Stizenb., op. cit. 81: 129. 1895. Bailey (1891a); Muell.-Arg. (1888a, 1891b). Q.

*S. INSCULPTA f. SUBLAEVIS (Muell.-Arg.) A. Zahlbr., Cat. Lich. Univ. 3: 388. 1925. Stictina impressula var. sublaevis Muell.-Arg., Hedwigia 30: 48. 1891. S. insculpta f. sublaevis Stizenberger, Flora 81: 129. 1895. Bailey (1891a); Shirley (1891b). Q.

S. LACERA (Hook. f. & Tayl.) Muell.-Arg., Flora 71: 130. 1888. Cetraria lacera Hook. f. & Tayl., London J. Bot. 3: 646. 1884. Sticta parvula Nyl., Lich. Nov. Zel. p. 33. 1888. Stirton (1899a). Q.

S. LATIFRONS Rich., Voy. Découvert. l'Astrolabe, Bot. 1: 27. 1832. F. Mueller (1881); Muell.-Arg. (1887c); Watts (1903); Zahlbruckner (1896). NSW.

S. LIMBATA Ach., Method. Lich. p. 280. 1803. Cheel (1912-14); Shirley (1889a); Wilson (1887). Q, V.

*S. LURIDOVIOLACEA (Stirt.) A. Zahlbr., Cat. Lich. Univ. 3: 393. 1925. Stictina luridoviolacea Stirt., Trans. New Zealand Inst. 32: 73. 1899. V. Cheel (1912-14).

S. MACROPHYLLA Bory ex Del., Hist. Lich. Sticta p. 110. 1822. Stictina macrophylla Nyl., Flora 52: 111. 1869. Cheel (1912-14); Shirley (1889a). Q.

S. MACROPHYLLA f. SPEIROCARPA Hue, Nouv. Arch. Mus. Paris (4) 3: 98. 1901. Stictina macrophylla var. speirocarpa Nyl., Flora 52: 118. 1869. Hue (1890-92, as "macrocarpa").

S. MARGINIFERA Mont., Ann. Sci. Nat., Bot. (2) 18: 265. 1842. Stictina marginifera Nyl., Bull. Soc. Linn. Normand. (2) 2: 53. 1868. Bailey (1881, 1883); Shirley (1889a); Wilson (1888, 1889b). NSW, Q, V.

*S. MOOREANA A. Zahlbr., Ann K.K. Naturhist. Hofmus. 11: 192. 1892. NSW.

S. MULTIFIDA Laur. ex Kremp., Verh. zool.-bot. Ges. Wien 18: 319. 1868. Hue (1890-92); Krempelhuber (1868, 1880); F. Mueller (1881); Muell.-Arg. (1887c); Stizenberger (1895). Sine loc.

S. ORBICULARIS (A. Br. ex Mey. & Flot.) Hue, Ann. Jard. Bot. Buitenzorg 17: 193. 1901. Sticta filicina var. orbicularis A. Br. ex Mey. & Flot., Nova Actorum Acad. Caes. Leop.-Carol. Nat. Cur. 19, Suppl., p. 215. 1843. Krempelhuber (1868). Sine loc.

S. PEDUNCULATA Kremp., J. Mus. Godeffroy 1(4): 97. 1874. *Sticta shirleyana Muell.-Arg., Hedwigia 32: 122. 1893. Shirley (1894). V.

S. PHYSCOSPOROIDES F. Mueller & Muell.-Arg., Fragm. Phytogeogr. Austral. 11, Suppl. 5: 116. 1881. Nomen nudum. Bailey (1883). Q.

*S. POCULIFERA Muell.-Arg., Flora 65: 304. 1882. Stizenberger (1895). NSW.

*S. PODOCARPA Muell.-Arg., op. cit. 74: 375. 1891. Cheel (1912-14); Stizenberger (1895). NSW.

S. PSILOPHYLLA Muell.-Arg., Bull. Soc. Bot. Belg. 31: 29. 1892. S. dissimulata var. multifida Nyl., Syn. Lich. 1: 363. 1860. Krempelhuber (1880); Zahlbruckner (1896). NSW.

S. PULVINATA (Mey. & Flot.) Vain., Philipp. J. Sci. C, 8: 123. 1913. S. filicina var. obicularis var. pulvinata Mey. & Flot., Nova Actorum Acad. Caes. Leop.-Carol. Nat. Cur. 19: 215. 1843. S. carpolomoides Nyl., Syn. Lich. 1: 354. 1860. Muell.-Arg. (1891a); Shirley (1891a). Q.

*S. PURPURASCENS (Stirt.) A. Zahlbr., Cat. Lich. Univ. 3: 362. 1925. Parmosticta purpurascens Stirt., Trans. Proc. New Zealand Inst. 32: 71. 1899. Sine loc.

*(STICTINA) PUSTULOSA F. Wils., Victoria Naturalist 6:60. 1889. Nomen nudum.

*S. RIGIDA (Muell.-Arg.) A. Zahlbr., Cat. Lich. Univ. 3: 398. 1925. Stictina rigida Muell.-Arg., Bull. Herb. Boiss. 4: 89. 1896. Muell.-Arg. (1898). Q.

*S. RUBRINA (Stirt.) Muell.-Arg., Flora 66: 23. 1883. Parmosticta rubrina Stirt., Trans. Proc. Roy. Soc. Queensland 17: 69. 1881. Bailey (1881, 1883); F. Mueller (1881). Q.

*S. RUTILANS (Stirt.) A. Zahlbr., Cat. Lich. Univ. 3: 398. 1925. Stictina rutilans Stirt., Proc. Roy. Soc. Victoria 17: 68. 1881. Bailey (1883); F. Mueller (1881); Shirley (1889a). Q.

*S. SAYERI Muell.-Arg., Flora 71: 28. 1888. Bailey (1891a); Bibby (1954); Muell.-Arg. (1891b). Q.

S. SEEMANII Bab. in Seem., Bot. Voy. "Herald" p. 248. 1852-57. Shirley (1893a). Q.

S. SINUOSA Pers. var. MACROPHYLLA Muell.-Arg., Hedwigia 30: 238. 1891. Sticta damaecornis var. macrophylla Bab. in Seem., Bot. Voy. "Herald" p. 247. 1852-57. Zahlbruckner (1896). NSW.

*S. STIPITATA Knight ex Wils., Pap. Proc. Roy. Soc. Tasmania (1892): 167. 1893. Wilson (1887, 1891a). V. =S. pedunculata fide Shirley, (1894.

S. SUBCAPERATA (Nyl.) Nyl., Lich. Nov. Zel. p. 31. 1888. Sticta damaecornis var. subcaperata Nyl., J. Linn. Soc. London, Bot. 9: 247. 1867. Cheel (1912-14). V.

S. SUBCORIACEA Nyl., J. Linn. Soc. London, Bot. 9: 247. 1865. Krempelhuber (1880); F. Mueller (1881). Sine loc.

*S. SUBCROCEA (Stirt.) A. Zahlbr., Cat. Lich. Univ. 3: 399. 1925. Stictina subcrocea Stirt., Trans. Proc. New Zealand Inst. 32: 74. 1899. Q.

*S. SUBERECTA (Stirt.) A. Zahlbr., Cat. Lich. Univ. 3: 399. 1925. Stictina suberecta Stirt., Trans. Proc. New Zealand Inst. 32: 73. 1899. Q.

*S. SUBORBICULARIS (Muell.-Arg.) A. Zahlbr., Cat. Lich. Univ. 3: 399. 1925. Stictina suborbicularis Muell.-Arg., Nuovo Giorn. Bot. Ital. 22: 387. 1891. Bailey (1891); Shirley (1891b, 1893a); Stizenberger (1895). Q.

*S. SUBTOMENTELLA (Knight ex Shirley) A. Zahlbr., Cat. Lich. Univ. 3: 399. 1925. Stictina subtomentella Knight ex Shirley, Proc. Roy. Soc. Queensland 6: 24. 1889. Q.

S. TOMENTELLA Nyl., Ann. Sci. Nat., Bot. (4) 11: 214. 1859. Stictina tomentella Nyl., Syn. Lich. 1: 342. 1860. Bailey (1881, 1883). F. Mueller (1881); Shirley (1889a). Q.

S. VARIABILIS (Bory) Ach. Lichenogr. Univ. p. 455. 1810. Lichen variabilis Bory, Voy. Quatr. Iles d'Afr. 1: 393. 1804. Bailey (1883); Cheel (1912-14); Hue (1890-92); Knight (1871); Krempelhuber (1880); F. Mueller (1881); Nylander (1857); Shirley (1889a, 1893a). NSW, Q, V.

S. WEIGELII Isert ex Ach. var. BEAUVOISII (Del.) Hue, Nouv. Arch. Mus. Paris (4) 3: 96. 1901. S. beauvoisii Del., Hist. Lich. Sticta p. 83. 1822. Cheel (1912-14). NSW.

*S. WEIGELII var. MICROPHYLLA (Kremp.) A. Zahlbr., Cat. Lich. Univ. 3: 406. 1925. Sticta quercizans var. microphylla Kremp., Verh. zool.-bot. Ges. Wien 30: 335. 1880. NSW.

STRIGULA

S. COMPLANATA (Fée) Nyl. var. CILIATA (Mont.) Muell.-Arg., Bot. Jahrb. 6: 380. 1885. Strigula ciliata Mont., Ann. Sci. Nat., Bot. (3) 10: 131. 1848. Bailey (1883); Muell.-Arg. (1895b); Shirley (1888b, 1889d). Q.

S. ELEGANS (Fée) Muell.-Arg., Linnaea 43: 41. 1880. Phyllocharis elegans Fée, Essai Crypt. Ecorc. Officin. p. xciv. 1824. Strigula plana Muell.-Arg., Bot. Jahrb. 6: 381. 1885. Bailey (1881, as S. complanata, fide Santesson 1952); F. Mueller (1881, as S. complanata); Muell.-Arg. (1895b); Santesson (1952). NSW, Q.

S. ELEGANS var. ANTILLARUM (Fée) R. Sant., Symb. Bot. Upsal. 12(1): 172. 1952. Melanophthalmum antillarum Fée, Essai Crypt. Ecorc. Officin. p. xciv and c. 1824. Q.

*S. ELEGANS var. ELATIOR (Stirt.) A. Zahlbr., Cat. Lich. Univ. 1: 539. 1922. S. elatior Stirt., Trans. Proc. Roy. Soc. Victoria 17: 75. 1881. Q. S. elegans var. eumorpha Muell.-Arg., Flora 68: 342. 1885. Bailey (1881, 1883); F. Mueller (1881); Muell.-Arg. (1895b); Shirley (1890, 1893a).

*S. ELEGANS var. PERTENUIS Muell.-Arg., Hedwigia 32: 134. 1893. Muell.-Arg. (1893b, 1895b). Q.

S. GLAZIOVII Muell.-Arg., Flora 73: 199. 1890. Muell.-Arg. (1891a, 1895b); Shirley (1891a). Q.

S. MACULATA (Cooke & Massee) R. Sant., Symb. Bot. Upsal. 12(1): 186. 1952. Micropeltis maculata Cooke & Massee in Cooke, Grevillea 18: 35. 1889. Santesson (1952). Q.

S. NEMATHORA Mont. in Sagra, Hist. l'Ile Cuba 9(2): 143. 1845. Bailey (1881, 1883); Knight (1884b); F. Mueller (1881); Muell.-Arg. (1895b); Santesson (1952). Q.

S. NITIDULA Mont. in Sagra, loc. cit., p. 139. Santesson (1952). Q.

S. SUBTILISSIMA (Fée) Muell.-Arg., Flora 66: 346. 1883. Racoplaca subtilissima Fée, Essai Crypt. Ecorc. Officin. p. xciv. 1824. Santesson (1952). NSW, Q.

SYNALISSA

*S. CANCELLATA F. Wils., Proc. Roy. Soc. Victoria n.s. 5: 151. 1892. Wilson (1887, 1891b, as S. micrococca Born. & Nyl.). V.

TAPELLARIA

T. PHYLLOPHILA (Stirt.) R. Sant. ex Thorold, J. Ecol. 40: 129. 1952. Lecidea phyllophila Stirt., Proc. Philos. Soc. Glasgow 10: 298. 1877. Santesson (1952). NSW.

TELOSCHISTES

T. CHRYSOPHTHALMUS (L.) Beltr., Lich. Bassan., p. 109. 1858. Lichen chrysophthalmus L., Mantissa p. 311. 1771. Physcia chrysophthalma DC. in Lam. & DC., Flore Franç. ed. 3, 2: 401. 1805. Bailey (1881, 1883); Bibby & Smith (1954); Filson (1969); Hellbom (1896); Hue (1890-92); Krempelhuber (1880); F. Mueller (1881); Muell.-Arg. (1883a, 1891a); Nylander (1857); Persoon (1826); Shirley (1889a); Stirton (1899b); Tate (1881, 1887); Turner (1905); Watts (1903); Willis (1953); Wilson (1889c). Q, V, SA, NSW, WA.

*T. CHRYSOPHTHALMUS var. ALATUS Shirley, Proc. Roy. Soc. Queensland 8: 133. 1892. Physcia comosa var. alata F. Wils. ex Bailey, Queensland Dept. Agr. Bull. 7: 32. 1891. Shirley (1892a,b). Q.

T. CHRYSOPHTHALMUS f. DENUDATUS (Hoffm.) Muell.-Arg., Flora 60: 265. 1883. Platysma denudatum Hoffm., Descr. Adumbr. Pl. Lich. 2: 23. 1794. Physcia chrysophthalma var. denudata Chév., Flore Génér. Env. Paris 1: 610. 1826. Stirton (1899b). Q.

*T. CHRYSOPHTHALMUS var. EXPALLENS Muell.-Arg., Flora 66: 78. 1883. NSW.

*T. CHRYSOPHTHALMUS var. LEUCOBLEPHARIS Muell.-Arg., loc. cit. p. 77. Physcia chrysophthalma var. leucoblephara Shirley, Lich. Fl. Queensland 4: 193. 1890. Shirley (1889b, 1892b). Q.

T. FASCICULATUS Hillm., Fedde Repert. 49: 176. 1938. Filson (1969); Weber (1969). NSW, V.

(Teloschistes)

T. FLAVICANS (Sw.) Norm., Nyt. Mag. Naturvidensk. 7: 229. 1853. Lichen flavicans Sw., Nova Gen. Sp. Pl. p. 147. 1788. Physcia flavicans DC. in Lam. & DC., Flore Franç. ed. 3, 2: 189. 1805. Bailey (1883); Filson (1969); Hue (1890-92); Krempelhuber (1873, 1880); F. Mueller (1881); Nylander (1857); Shirley (1889a). NSW, Q.

T. FLAVICANS var. COMPRESSUS Murray, Trans. Roy. Soc. New Zealand 88: 206. 1960. Murray (1960b). Sine loc.

T. FLAVICANS var. CROCEUS (Ach.) Muell.-Arg., Flora 71: 493. 1888. Cornicularia crocea Ach., Lichenogr. Univ. p. 615. 1810. Muell.-Arg. (1891a); Shirley (1891a). Q.

T. FLAVICANS var. SUBEXILIS (Nyl.) F. Wils., Pap. Proc. Roy. Soc. Tasmania for 1892, p. 176. 1893. Physcia subexilis Nyl. ex Cromb., J. Linn. Soc. London, Bot. 16: 395. 1879. Hue (1890-92). Sine loc. Tasmania only?

T. HYPOGLAUCUS (Nyl.) A. Zahlbr. in Engler & Prantl, Nat. Pflanzenfam. I. Teil, Abt. I*; 230. 1907. Physcia hypoglauca Nyl., Syn. Lich. 1: 409. 1860. F. Mueller (1881). Q.

*T. SIEBERIANUS (Laur.) Hillm., Hedwigia 69: 315. 1930. T. chrysophthalmus var. sieberianus Muell.-Arg., Flora 66: 77. 1883. Parmelia sieberiana Laur., Linnaea 2: 38. 1827. Physcia chrysopthalma var. sieberiana Shirley, Lich. Fl. Queensland 4: 193. 1890. Teloschistes sieberianus Hillm., Hedwigia 69: 315. 1930. Filson (1969); Muell.-Arg. (1883a, 1887c, 1892b); Shirley (1889b, 1892b); Stirton (1899b). NSW, SA, V, WA.

T. SPINOSUS (Hook. f. & Tayl.) J. Murray, Trans. Roy. Soc. New Zealand 88: 205. 1960. Parmelia spinosa Hook. f. & Tayl., London J. Bot. 3: 644. 1844. Xanthoria parietina var. spinulosa Muell.-Arg., Bull. Herb. Boiss. 2, Appendix 1: 40. 1894. X. spinosa DuRietz, Bot. Notiser p. 211. 1922. Physcia parietina var. spinulosa Kremp., Verh. zool.-bot. Ges. Wien 18: 322. 1868. Filson (1969); Hampe (1853); Krempelhuber (1880); Murray (1960b); Shirley (1894); Wilson (1889a, 1890a). V.

T. SPINOSUS f. SUBTERES Filson, Muelleria 2: 79. 1969. NSW, V.

*T. VELIFER F. Wils., Victoria Naturalist 6: 69. 1889. *T. chrysophthalmus var. fornicatus Muell.-Arg., Bull. Herb. Boiss. 4: 89. 1896. Filson (1969); Hampe (1853); Krempelhuber (1880); Muell.-Arg. (1898); Shirley (1894); Watts (1903); Weber (1969); Wilson (1889a, 1890a). NSW, V.

T. VELIFER f. NODULOSA (J. Murray) Filson, Muelleria 2: 81. 1969. T. fasciculatus var. nodulosa J. Murray, Trans. Roy. Soc. New Zealand 88: 206. 1960. Filson (1969). NSW, V.

T. XANTHORIOIDES J. Murray, Trans. Roy. Soc. New Zealand 88: 209. 1960. Filson (1969). NSW.

THAMNOLIA

T. VERMICULARIS (Sw.) ex Schaer., Enum. Crit. Lich. Europ. p. 243. 1850. Lichen vermicularis Sw., Method. Muscor. p. 37. 1781. Darbishire (1912); Hampe (1853); Hellbom (1896); Hue (1890-92); F. Mueller (1881); Shirley (1892a,b); Weber (1969). NSW, Q, V.

T. VERMICULARIS var. TAURICA (Wulf.) Schaer., Enum. Crit. Lich. Europ. p. 244. 1850. Lichen tauricus Wulf. in Jacq., Collect. Botan. 2: 177. 1788. Motyka (1960). NSW.

THELOTREMA

*T. ARGENTEUM Muell.-Arg., Hedwigia 30: 50. 1891. Bailey (1891a); Shirley (1891b). Q.

*T. AUSTRALIENSE Muell.-Arg., Flora 70: 61. 1887. (including Australian reports of T. microporellum Nyl., fide Muell.-Arg., 1887c). Bailey (1883); Kremp. (1880); Muell.-Arg. (1881, 1887, 1887c); Shirley (1889a, 1889b). Q.

*T. BICUSPIDATUM Muell.-Arg., Nuovo Giorn. Bot. Ital. 23: 395. 1891. Shirley (1891b). Q.

*T. CUPULARE Muell.-Arg., Hedwigia 32: 131. 1893. Shirley (1893a). Q.

*T. CYPHELLOIDES Muell.-Arg., Bull. Herb. Boiss. 3: 314. 1895. Shirley (1896). Q.

*T. DECORTICANS Muell.-Arg., op. cit. 1: 54. 1893. V.

T. DEPRESSUM Mont., Ann. Sci. Nat., Bot. (3) 16: 73. 1951. Ascidium depressum Nyl., Flora 52: 121. 1869. Bailey (1881, 1883); Muell.-Arg. (1881); Shirley (1889a). Q.

*T. ENDOXANTHUM Muell.-Arg., Nuovo Giorn. Bot. Ital. 23: 396. 1891. Shirley (1891b). Q.

*T. HYPOMELAENUM Muell.-Arg., Bull. Herb. Boiss. 3: 314. 1895. NSW.

*T. INTURGESCENS Muell.-Arg., Hedwigia 32: 131. 1893. Shirley (1893a). Q.

*T. LACERATULUM Muell.-Arg., Flora 70: 399. 1887. Shirley (1889a, 1889b). Q.

T. LEPADINUM (Ach.) Ach., Method. Lich., p. 132. 1803. Lichen lepadinus Ach., Lichenogr. Suec. Prodrom., p. 30. 1798. Bibby & Smith (1954); Muell.-Arg. (1892a, 1893); Wilson (1887). NSW, V, WA.

*T. "LEUCOPTHALMUM" Nyl. var. LACERATA Raes., Arch. Soc. Zool. Bot. Fenn. "Vanamo" 3: 184. 1949. This specific epithet was not published by Nylander, and it is impossible to know whether the error was for T. leucomelanum Nyl. without study of the specimen. NSW.

*T. MEGALOPHTHALMUM Muell.-Arg., Flora 65: 500. 1882. Knight (1884b); Muell.-Arg. (1882b, 1892a); Shirley (1889a). Q.

*T. MEGALOSPORUM Muell.-Arg., Nuovo Giorn. Bot. Ital. 23: 395. 1891. Shirley (1891b). Q.

*T. MICROPHTHALMUM Muell.-Arg., Bull. Herb. Boiss. 3: 314. 1895. Shirley (1896). Q.

*T. OBCONICUM Raes., Arch. Soc. Zool. Bot. Fenn. "Vanamo" 3: 184. 1949. NSW.

*T. PROFUNDUM Shirley, Proc. Roy. Soc. Queensland 6: 191. 1889. Ascidium profundum Stirton, Trans. Proc. Roy. Soc. Victoria 17: 70. 1881. Bailey (1881, 1883); Muell.-Arg. (1881, as A. "profusum," 1895c); Shirley (1889a, 1896). Q.

*T. RIMULOSUM Muell.-Arg., Nuovo Giorn. Bot. Ital. 23: 396. 1891. Shirley (1891b). Q.

THYSANOTHECIUM

*T. HOOKERI Mont. & Berk., London J. Bot. 5: 257. 1846. WA. Bibby & Smith (1954); Hue (1890-92); Nylander (1857); Stirton (1899b); Willis (1953); Wilson (1892). Q, V, WA.

*T. HYALINUM (Tayl.) Nyl., Mém. Soc. Scienc. Nat. Cherbourg 5: 94. 1857. Baeomyces hyalinus Tayl., London J. Bot. 6: 187. 1847. WA. *Cladonia? scutellata Fr., Plantae Preiss. p. 141. 1846-47. Sine loc. Hue (1890-92); Krempelhuber (1880); Leighton (1867); F. Mueller (1881); Muell.-Arg. (1888a); Nylander (1857); Shirley (1889a); Tate (1882); Weber (1969); Wilson (1887, 1891b, 1892). Q, SA, V, WA.

*T. HYALINUM f. INTORTUM F. Wils., Proc. Roy. Soc. Victoria n.s. 5: 176. 1892. V.

TOMASELLIA

T. ACICULIFERA (Nyl.) Muell.-Arg., Bot. Jahrb. 5: 398. 1885. Melanotheca aciculifera Nyl., Expos. Synopt. Pyrenocarp. p. 71. 1858. Bailey (1891a); Muell.-Arg. (1891b, 1895b); Shirley (1891a). Q.

*T. DISPORA Muell.-Arg., Flora 70: 427. 1887. Muell.-Arg. (1895b); Shirley (1889b, 1890). Q.

T. GELATINOSA (Chev.) A. Zahlbr., Cat. Lich. Univ. 1: 474. 1922. Hue (1890-92); Nylander (1886). NSW.

*T. QUEENSLANDICA Muell.-Arg., Rep. Austral. Assoc. Adv. Sci. p. 460. 1895. Shirley (1895). Q.

TONINIA

*T. AUSTRALIENSIS (Muell.-Arg.) A. Zahlbr., Flora 70: 320. 1887. Thalloidima australiense Muell.-Arg., Hedwigia 5: 195. 1892. WA. Shirley (1894). V, WA.

T. BULLATA (Mey. & Flot.) A. Zahlbr., Bot. Centralbl. Beih. 19, 2. Abt.: 76. 1905. Lecidea bullata Mey. & Flot., Nova Actorum Acad. Caes. Leop.-Carol. Nat. Cur. 19 (Suppl.): 227. 1843. *Siphula muelleri F. Wils., Victoria Naturalist 6: 179. 1890. V. Bibbya muelleri J. H. Willis, op. cit. 73: 125. 1956. Muell.-Arg. (1898); Weber (1969). NSW, V.

T. COERULEONIGRICANS (Lightf.) Th. Fr., Lichenogr. Scand. 1: 336. 1874. Lichen coeruleonigricans Lightf., Fl. Scotica 2: 805. 1777. Muell.-Arg. (1893); Shirley (1891b, 1893a). Q, V.

*T. CONGLOMERANS (Muell.-Arg.) A. Zahlbr., Cat. Lich. Univ. 4: 285. 1926. Thalloidima conglomerans Muell.-Arg., Bull. Herb. Boiss. 1: 36. 1893. V.

*T. LEUCINA (Muell.-Arg.) A. Zahlbr., Cat. Lich. Univ. 4: 289. 1926. Thalloidima leucinum Muell.-Arg., Bull. Herb. Boiss. 1: 35. 1893. V.

*T. MICROLEPIS (Muell.-Arg.) A. Zahlbr., Cat. Lich. Univ. 4: 275. 1926. Thalloidima microlepis Muell.-Arg., Bull. Herb. Boiss. 1: 35. 1893. V.

*T. NITIDA (Muell.-Arg.) A. Zahlbr., Cat. Lich. Univ. 4: 290. 1926. *Thalloidima nitidum Muell.-Arg., Bull. Herb. Boiss. 4: 92. 1896. Lamb (1954); Muell.-Arg. (1898). V.

TRAPELIA

T. COARCTATA (Turn. in Sm. & Sowerby) Choisy in Wern., Bull. Soc. Sci. Nat. Maroc 12: 160. 1932. Lichen coarctatus Turn. in Sm. & Sowerby,

Engl. Bot. 8: tab. 534. 1799. Lecidea coarctata Nyl., Act. Soc. Linn. Bordeaux 21: 358. 1856. Lecanora coarctata Ach., Lichenogr. Univ. p. 352. 1810. Hampe (1853); F. Mueller (1881); Nylander (1857). V.

(LECIDEA) COARCTATA var. ELACISTA (Ach.) Th. Fr., Lichenogr. Scand. 1: 447. 1871. Parmelia elacista Ach., Method. Lich. p. 159. 1803. Muell.-Arg. (1893). V.

(BIATORA) COARCTATA f. TERRESTRIS Flot. in Cohn, Krypt.-Fl. Schles. 2 (2): 194. 1879. Wilson (1887). V.

TREMOTYLIUM

*T. AUSTRALIENSE Muell.-Arg., Flora 65: 500. 1882. Shirley (1889a). Q.

*T. NITIDULUM Muell.-Arg., Hedwigia 32: 132. 1893 (listed, incorrectly, as T. nitidum by Zahlbr., Cat. Lich. Univ. 2: 645. 1924). Shirley (1893a). Q.

TRICHARIA

T. VAINIOI R. Sant., Symb. Bot. Upsal. 12(1): 382. 1952. Q.

TRICHOTHELIUM

T. ALBOATRUM Vain., Ann. Acad. Sci. Fenn. ser. A., 15: 321. 1921. Santesson (1952). Q.

TRYPETHELIUM

T. ANOMALUM Ach., Syn. Lich. p. 105. 1814. T. platystomum Mont., Ann. Sci. Nat., Bot. (12) 19: 72. 1843. Harmand (1911); Muell.-Arg. (1895b); Shirley (1892a). Q.

T. CATERVARIUM (Fée) Tuck., Gen. Lich. p. 260. 1872. Verrucaria catervaria Fée, Essai Crypt. Ecorc. Officin. p. 90. 1824. Bailey (1881, 1883); F. Mueller (1881); Muell.-Arg. (1895b). Q.

T. ELUTERIAE Spreng., Einl. Stud. Krypt.-Gew. p. 351. 1804. T. sprengelii Ach., Lichenogr. Univ. p. 306. 1810. Bailey (1881, 1883); Harmand (1911); F. Mueller (1881); Muell.-Arg. (1891a, 1895b). Q.

T. ELUTERIAE var. CITRINUM (Eschw.) Muell.-Arg., Bot. Jahrb. 6: 393. 1885. Astrothelium varium, citrinum Eschw. in Mart., Fl. Brasil. 1: 162. 1833. Muell.-Arg. (1891a, 1895b); Shirley (1891a). Q.

*T. EXIGUELLUM Stirt. ex Bailey, Queensland Agr. J. 5: 39. 1899. Q.

T. FUMOSO-CINEREUM F. Wils., Victorian Naturalist 5: 31. 1888. Nomen nudum. Wilson (1889b,c). NSW, V.

T. INFUSCATULUM Muell.-Arg., Bot. Jahrb. 6: 389. 1885. Muell.-Arg. (1891b, 1895b); Shirley (1895). Q.

T. MASTOIDEUM (Ach.) Ach., Lichenogr. Univ. p. 307. 1810. Bathelium mastoideum Ach., Method. Lich. p. 111. 1803. Muell.-Arg. (1891a, 1895b); Shirley (1890). Q.

*T. MEDIANS Harm., Bull. Soc. Sci. Nancy (3) 12: 142. 1911. Sine loc.

*T. OLIGOCARPUM Muell.-Arg., Nuovo Giorn. Bot. Ital. 23: 402. 1881. Muell.-Arg. (1895a); Shirley (1895). Q.

T. PAPILLOSUM Ach., Syn. Lich. p. 104. 1814. Bailey (1881, 1883); F. Mueller (1881, 1895b). Q.

T. TROPICUM (Mass.) Muell.-Arg., Bot. Jahrb. 6: 393. 1885. Sagedia tropica Mass., Ricerch. Auton. Lich. p. 161. 1852. Verrucaria nitidiuscula Nyl., Acta Soc. Sci. Fenn. 7: 491. 1863. Bailey (1881, 1883, 1891a); Harmand (1911); F. Mueller (1881); Muell-Arg. (1891a, b, 1895b). Q.

*T. TROPICUM var. NIGRATUM Muell.-Arg., Rep. Austral. Assoc. Adv. Sci. p. 460. 1895 (overlooked by Zahlbruckner, Cat. Lich. Univ.) Shirley (1895). Q.

*T. VIRGINEUM Muell.-Arg., loc. cit. p. 461. Shirley (1895). Erroneously listed as T. "virginicum" by Zahlbruckner, Cat. Lich. Univ. 1: 502. 1922. Q.

TYLOPHORON

*T. TRILOCULARE Muell.-Arg., Hedwigia 32: 122. 1893. Shirley (1893a). Q.

UMBILICARIA

U. CYLINDRICA (L.) Del. in Duby, Bot. Gall. 2: 595. 1830. Lichen cylindricus L., Sp. Pl. p. 1144. 1753. Hampe (1853); Hue (1890-92); Llano (1950); Weber (1969); Wilson (1890a). V.

U. DECUSSATA (Vill.) A. Zahlbr., Cat. Lich. Univ. 8: 490. 1942. Lichen decussatus Vill., Hist. Plant. Dauphiné 3: 964. 1789. Llano (1950); Weber (1969). V.

U. HIRSUTA (Sw.) Ach., Kongl. Svenska Vetensk.-Akad. Nya Handl. 15: 97. 1794. Lichen hirsutus Sw. ex Westr., op. cit. 14: 47. 1793. A.C.T.: summit of Mt. Coree, under rock overhang, Dec. 1965, D. McVean 65133 (COLO).

U. POLYPHYLLA (L.) Baumg., Fl. Lips. p. 571. 1790. Lichen polyphyllus L., Sp. Pl. p. 1150. 1753. Llano (1950); Wilson (1890a, as f. anthracina Ach.). V.

U. PROBOSCIDEA (L.) Schrad., Spicil. Fl. German 1: 103. 1794. Lichen proboscideus L., Sp. Pl. p. 1150. 1753. Llano (1950); Weber (1969).

U. SUBGLABRA (Nyl.) Harm., Lich. France 4: 707. 1909. Gyrophora subglabra Nyl., Lich. Env. Paris p. 135. 1896. Agyrophora subglabra Llano, Monogr. Umbil. p. 69. 1950. Llano (1950); Weber (1969). NSW, V.

USNEA

*U. ANGULOSA (Muell.-Arg.) Mot., Monogr. Usnea p. 512. 1936-38. U. dasypogoides var. angulosa Muell.-Arg., Flora 69: 254. 1886. WA.

*U. ARIDA Mot., Monogr. Usnea p. 492 1936-38. V. Also in NSW, SA.

U. ARTICULATA (L.) Hoffm., Deutschl. Fl. 2: 133. 1796. Lichen articulatus L., Sp. Pl. p. 1156. 1753. Bailey (1881, 1883); Hellbom (1896); Shirley (1889a); Wilson (1887). Q. V. Motyka does not list for Australia.

U. ARTICULATA var. INTESTINIFORMIS (Ach.) Cromb., Grevillea 15: 48. 1886. U. barbata var. intestiniformis Ach., Lichenogr. Univ. p. 625. 1810. Krempelhuber (1880); Muell.-Arg. (1887c). Sine loc. Motyka does not list for Australia.

*U. BAILEYI (Stirt.) A. Zahlbr., Denkschr. Kaiserl. Akad. Wiss., Math.-Naturw. Kl. 83: 182. 1909. Eumitria baileyi Stirt., Scott. Naturalist 6: 100. 1881. NSW. Usnea barbata var. asperrima Muell.-Arg., Flora 65: 299. 1882. Bailey (1883); F. Mueller (1881); Muell.-Arg. (1891a); Shirley (1889a, 1891a, 1893a); Wilson (1889b). NSW, Q.

(Usnea)

U. BARBATA (L.) Wigg. emend. Mot., Monogr. Usnea p. 206. 1936-38. U. barbata Wigg., Primit. Fl. Holsat. p. 91. 1780. Lichen barbatus L., Sp. Pl. p. 1155. 1793. Bailey (1883); Bibby & Smith (1954); Fries (1846-47); F. Mueller (1881); Shirley (1889a, 1892b); Tate (1881, 1887); Turner (1905); Wilson (1889c, 1890a). NSW, Q, SA, V, WA. Motyka does not list for Australia.

U. BARBATA var. CORNUTA Flot., Linnaea 17: 16. 1843. (Usnea cornuta of Australian reports.) Bailey (1881); Shirley (1889a). It is highly doubtful to which species this report refers.

(U. BARBATA var. MICROCARPA Pers. Wilson [1887]. We are unable to find the citation for this epithet). V.

U. CAVERNOSA Tuck. in Agassiz, Lake Superior, p. 171. 1850. Darbishire (1912). Sin loc. Undoubtedly a misidentification.

U. CERATINA Ach., Lichenogr. Univ. p. 619. 1810. Crombie (1880); Krempelhuber (1868); Maiden (1898); F. Mueller (1881); Shirley (1889d); Stirton (1899b); Watts (1903); Zahlbruckner (1896). NSW, Q, V. Motyka does not list for Australia.

U. CONTORTA Jatta, Malpighia 19: 163. 1905. Motyka (1936-38). NSW.

U. DASYPOGA (Ach.) Roehl. emend. Mot., Monogr. Usnea p. 189. 1936-38. U. dasypoga Roehl., Deutschl. Fl. 3(2): 144. 1813. Krempelhuber (1868); Muell.-Arg. (1891a); Shirley (1889a, 1892b); Watts (1903). Q. Motyka does not list for Australia.

U. DASYPOGOIDES Nyl. ex Crombie, J. Bot. 14: 263. 1876. Shirley (1889a); Wilson (1887). Q, V. Although Motyka cites Shirley (1889a), he attributes this species only to Rodriguez I. and Juan Fernandez.

U. DICHOTOMA Fries, Syst. Orb. Veg. 1: 282. 1825. Watts (1903); Zahlbruckner (1896, as U. vrieseana Mont.). NSW. (not verified by Motyka 1936-38).

U. FLORIDA (L.) Wigg., Primit. Fl. Holsat. p. 91. 1780. Lichen floridus L., Sp. Pl. p. 1156. 1753. Crombie (1880); Hampe (1853); Hue (1890-92); Krempelhuber (1868); Laurer (1827); Shirley (1889a). Q, V. Motyka does not list for Australia.

*U. FORMOSA (Stirt.) A. Zahlbr., Cat. Lich. Univ. 6: 575. 1930. Eumitria formosa Stirt., Scott. Naturalist 6: 297. 1882. Motyka (1936-38). Q.

U. FUSCORUBENS Mot., Monogr. Usnea p. 546. 1936-38. Motyka (1936-38). Sine loc.

*U. GLOMERATA Mot., loc. cit., p. 315. Sine loc.

U. HAWAIIENSIS Mot., loc. cit., p. 502. U. barbata var. scabrosa Muell.-Arg., Jahrb. Koenigl. Bot. Gart. Berlin 2: 308. 1883. Wilson (1887). Motyka does not list for Australia.

*U. HIMANTHODES Stirt., Scott. Naturalist 7: 75. 1883. Motyka (1936-38); Stirton (1882, 1897, 1898). NSW.

U. HIRTA (L.) Wigg. emend. Mot., Monogr. Usnea. p. 83. 1936-38. U. hirta Wigg., Primit. Fl. Holsat. p. 91. 1780. Lichen hirtus L., Sp. Pl. p. 1155. 1753. U. florida var. hirta Ach., Method. Lich. p. 309. 1803. U. barbata var. hirta Fr., Lichenogr. Europ. Reform. p. 18. 1831. Hampe (1853); Muell.-Arg. (1891a); Shirley (1891a); Stirton (1899b); Watts (1903). Q, SA. Not allowed for Australia by Motyka. Reports probably refer to U. scabrida.

*U. IMPLICITA (Stirt.) A. Zahlbr., Cat. Lich. Univ. 6: 582. 1930. Eumitria implicita Stirt., Scott. Nat. 6: 100. 1881. *U. dasypogoides var. substrigosa Muell.-Arg., Flora 72: 143. 1889. Q. Motyka (1936-38); Wilson (1887). NSW, Q.

*U. INERMIS Mot., Monogr. Usnea p. 109. 1936-38. V. Also NSW, SA.

U. INFLATA Del. in Duby, Bot. Gall. 2: 615. 1830. U. ceratina var. scabrosa Ach., Lichenogr. Univ. p. 620. 1810. Laurer (1827). Motyka does not list for Australia.

U. INTERCALARIS Kremp., J. Mus. Godeffroy 1 (4): 96. 1874. Bailey (1883); Krempelhuber (1880); F. Mueller (1881); Shirley (1889a); Zahlbruckner (1896). None cited by Motyka.

U. LONGISSIMA Ach., Lichenogr. Univ. p. 626. 1810. Bailey (1881, 1883); Hue (1890-92); Krempelhuber (1880); F. Mueller (1881); Muell.-Arg. (1891a); Shirley (1889a); Stirton (1897, 1898, 1899b); Turner (1905); Wilson (1889b). None cited by Motyka. Possibly these reports may refer to the closely related U. trichodeoides Vain.

*U. LURIDORUFA Stirt. ssp. PALLIDA Stirt., Scott. Naturalist 6: 295. 1882. Q. Motyka does not appear to have disposed of this taxon.

U. MACULATA Stirt., Scott. Naturalist 6: 293. 1882. Motyka (1936-38). Sine loc.

*U. MICROCARPOIDES (Muell.-Arg.) Mot., Monogr. Usnea p. 525. 1936-38. U. dasypogoides var. microcarpoides Muell.-Arg., Flora 66: 20. 1883. Q. Weber (1971). NSW.

U. MISAMINENSIS (Vain.) Mot., Monogr. Usnea p. 418. 1936-38. Motyka (1936-38). NSW, Q.

*U. MOLLIUSCULA Stirt., Scott. Naturalist 7: 77. 1883. Motyka (1936-38); Stirton (1882, 1897, 1898). V.

U. PLICATA (L.) Wigg., Primit. Fl. Holsat. p. 91. 1780. Lichen plicatus L., Sp. Pl. p. 1154. 1753. Usnea barbata var. plicata Fr., Lichenogr. Europ. Reform. p. 18. 1830. Krempelhuber (1868); Laurer (1827); Wilson (1887). V. Motyka does not list for Australia.

U. POLIOTRIX Kremp., Vidensk. Meddel. Dansk Naturhist. Foren. Kjoebenhavn 5: 4. 1873. Watts (1903); Weber (1971). Motyka does not list for Australia.

*U. PROPINQUA (Stirt.) Stirt., Scott. Naturalist 7: 76. 1883. U. chaetophora Stirt. var. propinqua Stirt., Trans. Proc. New Zealand Inst. 30: 391. (1897) 1898. Motyka (1936-38). V.

*U. PULVINATA Fries in Lehmann, Plant. Preiss. 2: 145. 1846-47. U. tropica A. Zahlbr., Cat. Lich. Univ. 6: 596. 1930. *U. pectinata Stirt., Scott. Naturalist 7: 77. 1883 (non Tayl., 1847) V. Motyka (1936-38); Stirton (1897, 1898).

*U. QUEENSLANDICA Mot., Monogr. Usnea p. 612. 1936-38. Q, Sine loc.

U. RUBIGINEA (Michx.) Mass., Mem. Imp. Regia Instit. Veneto 10: 45. 1861. U. florida var. rubiginea Michx., Fl. Bor.-Am. 2: 332. 1803. U. rubescens Stirt., Scott. Naturalist 7: 76. 1883. U. sublurida Stirt., Trans. Proc. New Zealand Inst. 30: 389. (1897) 1898 (non Stirt., 1881). Bailey (1881); Motyka (1936-38); Stirton (1882, 1897, 1898, 1899b); Weber (1969); Wilson (1887). NSW, Q, V.

(Usnea)

*U. SCABRIDA Tayl., Phytologist 1: 1095. 1844. WA. U. barbata var. scabrida Muell.-Arg., Rev. Mycol. 1: 193. 1847. U. elegans Stirt., Trans. Proc. Roy. Soc. Victoria 17: 68. 1881. Q. U. dasypogoides var. elegans Muell.-Arg., Flora 66: 20. 1883 (pr. p.) Q. U. barbata var. elegans Muell.-Arg., op. cit. 72: 143. 1889 (pr. p.). *U. consimilis Stirt., Scott. Naturalist 6: 295. 1882. Sine loc. Bailey (1881, 1883); Bibby & Smith (1954); Motyka (1936-38); F. Mueller (1881); Muell.-Arg. (1883a, 1891a, 1892a, 1892b); Shirley (1889a, 1891a); Stirton (1881b, 1882, 1897, 1898, 1899b); Weber (1971); Wilson (1887); Zahlbruckner (1896). NSW, Q, SA, V. WA.

*U. SPILOTA Stirt., Scott. Naturalist 6: 294. 1882. V. Motyka (1936-38); Stirton (1882, 1897, 1898).

U. STRAMINEA Muell.-Arg. emend. Motyka, Monogr. Usnea p. 466. 1936-38. U. straminea Muell.-Arg., Flora 62: 162. 1879 (pr. p., excl. synon.). U. trichodea Shirley, Proc. Roy. Soc. Queensland 5: 105. 1888 (pr. p.). Motyka (1936-38). Q.

U. STRIGOSA (Ach.) Eaton, Man. Bot. ed. 5: 431. 1829. U. florida var. strigosa Ach., Method. Lich. p. 310. 1803. U. barbata var. strigosa Tuck., Proc. Amer. Acad. Arts Sci. 1: 200. 1848. Muell.-Arg. (1891a); Shirley (1891a); Wilson (1887). Q, V. Motyka does not list for Australia.

U. STRIGOSELLA Stein., Bull. Herb. Boiss. (2) 7: 637. 1907. Motyka (1936-38). NSW.

*U. SUBSORDIDA Stirt. var. TENEBROSA Stirt., Trans. New Zealand Inst. 30: 389. (1897) 1898. Q. Motyka does not dispose of this taxon.

*U. TORQUESCENS Stirt., Trans. New Zealand Inst. 30: 391. (1897) 1898. NSW. U. undulata Stirt., Scott. Naturalist 7: 75. 1883 (non Stirt., 1881). Usnea angulata of Australian reports, fide Motyka (1936-38). Hellbom (1896); Knight (1884b); Krempelhuber (1880); Motyka (1936-38); Muell.-Arg. (1891a); Shirley (1889a); Stirton (1897, 1898); Weber (1971). NSW, Q.

*U. TORULOSA (Muell.-Arg.) A. Zahlbr., Cat. Lich. Univ. 6: 594. 1930. U. dasypogoides var. torulosa Muell.-Arg., Flora 66: 19. 1883 (pr. p.). Motyka (1936-38); Weber (1971). NSW.

U. TRICHODEOIDES Vain. emend. Mot., Monogr. Usnea p. 421. 1936-38. (Usnea trichodea of some Australian reports, fide Motyka, loc. cit.): Bailey (1883); Hellbom (1896); Krempelhuber (1868, 1880); F. Mueller (1881); Shirley (1889a); Turner (1905); Watts (1903). NSW, Q, V.

U. UNDULATA Stirt. f. PERSPINIGERA Asah., J. Jap. Bot. 42: 325. 1967. NSW, Q.

U. XANTHOPOGA Nyl., Compt. Rend. Hébd. Séanc. Acad. Paris 83: 89. 1876. U. barbata var. xanthopoga Muell.-Arg., Flora 72: 143. 1899. Bibby & Smith (1954); Muell.-Arg. (1892b). WA. Motyka does not list for Australia.

VERRUCARIA

V. CALCISEDA DC. in Lam. & DC., Flore Franç. 2: 317. 1805. Muell.-Arg. (1892b). WA.

V. CEUTHOCARPA Wahlenb. ex Ach., Method. Lich. p. 22. 1803. Muell.-Arg. (1893). V.

V. MAURA Wahlenb. ex Ach., loc. cit. p. 19. 1803. Lichen maurus Sm. in Sm. & Sowerb., Engl. Bot. 35: tab. 2456. 1812. Muell.-Arg. (1893); Wilson (1887). V.

V. MUCOSA Wahlenb. ex Ach., Method. Lich. p. 23. 1803. Muell.-Arg. (1893). V.

V. RUPESTRIS Schrad., Spicil. Flor. German. 1: 109. 1794. V. muralis Ach., Method. Lich. p. 115. 1803. Muell.-Arg. (1893). V.

V. SPHINCTRINA (Duf.) Ach., Syn. Lich. p. 91. 1814. Limboria sphinctrina Duf. ex Fr., Lichenogr. Europ. Reform. p. 456. 1831. Muell.-Arg. (1893). V.

XANTHORIA

X. CANDELARIA (L.) Th. Fr. var. LACINIOSA (Duf.) Arn., Flora 67: 244. 1884. Parmelia parietina var. laciniosa Duf. ex Schaer., Enum. Critic. Lich. Europ. p. 51. 1850. Xanthoria controversa var. laciniosa Muell.-Arg., Hedwigia 31: 192. 1892. Muell.-Arg. (1892b). WA.

X. ECTANEA (Ach.) Raes. ex Filson, Muelleria 2: 83. 1969. Parmelia parietina var. ectanea Ach., Lich. Univ. 464. 1810. Xanthoria parietina var. ectanea (Ach.) Kickx., Fl. Cryptog. Flandres 2: 228. 1867. Filson (1969); Hue (1890-92); Nylander (1857). NSW, SA, V, WA.

X. PARIETINA (L.) Th. Fr., Nova Acta Reg. Soc. Scient. Upsal. (3) 3: 167. 1861. Lichen parietinus L., Sp. Pl. p. 1143. 1753. Parmelia parietina Ach., Method. Lich., p. 213. 1803. Physcia parietina DeNot., Giorn. Bot. Ital. 2 (2): 197. 1847. *Physcia ligulata Koerb., Abh. Schles. Ges. vaterl. Kultur 2: 30. 1862. Bibby & Smith (1954); Filson (1969); Fries (1846-47); Hue (1890-92); Krempelhuber (1880); F. Mueller (1881); Muell.-Arg. (1887c, 1892b); Nylander (1857); Tate (1881); Willis (1953). NSW, V. WA.

LICHEN PARASITES

ABROTHALLUS PARMELIARUM (Sommerf.) Arn., Flora 57: 102. 1874. Lecidea parmeliorum [sic] Sommerf., Suppl., Fl. Lapp. p. 176. 1826. Muell.-Arg. (1893a). V.

*ARTHONIA RICASOLIAE Muell.-Arg., Flora 70: 424. 1887. Q. Shirley (1889a,b).

*NESOLECHIA COCCOCARPIAE Muell.-Arg., Flora 70: 397. 1887. Q. Shirley (1889a,b).

*NESOLECHIA RUFA Muell.-Arg., Bull. Herb. Boiss. 1: 47. 1893. V.

OBRYZUM: records of this genus (=Guignardia: Sphaeriales) are assigned to non-lichenized fungi.

PATELLARIA LAURINEARUM F. Mueller, Fragm. Phytogeogr. Austral., XI. Suppl. p. 117. 1881. Nomen nudum.

*PATELLARIA RAMALINAE Muell.-Arg., Flora 66: 79. 1883. WA.

INDEX TO TAXONOMIC SYNONYMS

(Index to Taxonomic Synonyms Continued)

(Index to Taxonomic Synonyms Continued)

BIBLIOGRAPHY

Abbayes, H. Des. 1966. Sur quelques Cladonia (Lichens) exotiques, nouveaux ou peu connus. Rev. Bryol. Lichénol. 34: 821-828.

Acharius, E. 1814. Synopsis methodica lichenum, pp. 248, 343. (Cheel 101).

Ahti, Teuvo. 1961. Taxonomic studies on reindeer lichens (Cladonia, subgenus Cladina). Ann. Bot. Soc. Zool.-Bot. Fenn. "Vanamo" 32(1): 1-160. Pl. 1-44.

Anonymous. 1892. [List of lichen genera collected "up the river" in Queensland by the Field Naturalists during the year]. No species mentioned. Proc. Roy. Soc. Queensland 84 (4), appendix II: vi.

Bailey, F. M. 1881. The lichens of Queensland (with introduction by Rev. J. E. Tenison-Wood). Pap. & Proc. Roy. Soc. Tasmania for 1880-81, pp. 26-39. (Cheel 76 gives Tenison-Wood authorship).

__________ 1883. A synopsis of the Queensland flora. Brisbane (pp. 740-753 by C. Knight). (Cheel 68).

__________ 1884a, b, c. Contributions to the Queensland flora. Proc. Roy. Soc. Queensland 1: 70-78; 2: 84-92; 3: 106-113. (Cheel 69, 70, 71).

*__________ 1890. A synopsis of the Queensland flora, containing both Phanerogamous and Cryptogamous plants. 1st Supplement (1886); 2nd Supplement (1888); 3rd Supplement (1890). Brisbane.

__________ 1890a. Catalogue of the plants of Queensland. Brisbane. A secondary source, list only.

__________ 1890b. Contributions to the Queensland flora. Queensland Dept. Agr. Bull. 4 (Bailey's Bot. Bull. 1). No lichens listed.

__________ 1891a. Final Supplement to the report of the Bellenden Ker Expedition (determinations by Prof. J. Mueller). Queensland Dept. Agr. Ann. Rep. for 1890-91, pp. 43-44. A bare list without localities.

__________ 1891-96. Contributions to the Queensland flora. Queensland Dept. Agr. Bull. 7, 9, 13, 18 (Bailey's Bot. Bull. 2-5, 8, 10, 13). See also Shirley, and Wilson.

*__________ 1891. Presidential address, Concise history of Australian botany. Proc. Roy. Soc. Queensland 8(2): xvi-sli.

Baker, R. T. 1899. Contribution to a knowledge of the flora of Australia, No. 2. Proc. Linn. Soc. New South Wales 24: 446. (Cheel 16).

__________ 1902. Contribution to a knowledge of the flora of Australia, No. 4. op. cit. 27: 544. (Cheel 143).

Bennett, George. 1834. Wanderings in New South Wales, Batavia, Pedir Coast, Singapore, and China; being the journal of a naturalist in those countries during 1832, 1833, and 1834. London. 2 vols. A travel narrative with no significant lichen entries.

*Bennett, John Joseph. 1876. Robert Brown, Iter Austral. (exsiccati issued of Australian and Tasmanian lichens).

Bentham, George. 1863. Flora Australiensis. Lovell & Reeve & Co., London. 7 vol. No lichens mentioned, but much information on collectors and localities.

Bibby, P. N. S. 1942. Victorian lichens. Victoria Naturalist 59: 108-110. A general introduction, no species listed.

__________ 1946. Iceland moss: appears in Australia. <u>op</u>. <u>cit</u>. 63: 153. <u>Cetraria islandica</u> found in Victoria.

__________ 1954. Cryptogams of the 1948 Archbold Cape York (Queensland) Expedition. Jour. Arnold Arbor. 35(3): 260-265. Merely a list of collections by L. J. Brass, with localities.

__________ 1956. A remarkable lichen from arid Australia. Muelleria 1(1): 60. <u>Parmelia</u> (Chondropsis) <u>semiviridis</u>.

__________ 1959. Some lichens collected in Arnhem Land. P. 169 in Specht, R. L. and C. P. Mountford, Records of the American-Australian Scientific Expedition to Arnhem Land. 3. Botany and Plant Ecology. xv + 522 pp. illustr. Melbourne University Press.

Bibby, P. N. S. & G. G. Smith. 1954 (1955). List of lichens of Western Australia. J. Roy. Soc. W. Austral. 39: 28-29.

Bitter, G. 1901. Zur Morphologie und Systematik von <u>Parmelia</u>, Untergattung Hypogymnia. Hedwigia 40: 171-204. Tab. x, xi. (Cheel 144).

Brown, Robert. 1810. Prodromus florae Novae Hollandiae et insulae Van Diemen, exhibens characteres plantarum quas annis 1802-1805 ... Vol. I. London. (Cheel 102).

__________ 1814. Appendix, pp. 593-594, in Flinders, Matthew. A voyage to Terra Australis. A list of lichens "common to Europe and Terra Australis," a secondary source.

__________ 1814a. General remarks geographical and systematical on the botany of Terra Australis. London.

*Burnett, P. E. B. 1925. Coral lichen. Austral. Naturalist 5: 239. <u>Cladia retipora</u> in New South Wales.

Cheel, Edwin. 1901-03. Director's reports [by J. H. Maiden] to the Botanic Gardens, Sydney, presented to the legislative Assembly, New South Wales. (Cheel 17). Several collections mentioned from various parts of New South Wales.

__________ 1902 (1903). Bibliography of Australian lichens. Proc. Roy. Soc. New South Wales 33: xxxvii. A notice only.

__________ 1903. Bibliography of Australian lichens. J. & Proc. Roy. Soc. New South Wales 37: 172-182.

__________ 1903-1906. Director's report [by J. H. Maiden] on the Botanical Gardens, Sydney, presented to the legislative assembly, New South Wales, (Cheel 142). No significant entries.

__________ 1906. Bibliography of Australian, New Zealand, and South Sea Island lichens (second paper). J. Proc. Roy. Soc. New South Wales 40: 141-154. Several of the items are sloppily cited and as a result were difficult to locate, but these papers are the basic bibliography of Australian lichenology.

__________ 1912-13. Australian and South Sea Islands Stictaceae. I. Austral. Assoc. Adv. Sci. 13: 254-270 (1912); II. <u>Op</u>. <u>cit</u>. 14: 311-320.

Cheel, Edwin. 1924. Notes on a "coral lichen" (Cladonia retipora). Austral. Naturalist 5: 183-186.

Crombie, J. M. 1880. Enumeration of lichens in herb. Robert Brown (Brit. Mus.), with descritions of new species. J. Linn. Soc. London, Bot. 17: 390-401. (Cheel 3).

Darbishire, O. V. 1897. Ueber die Flechtentribus der Roccellei. Ber. Deutsch. Bot. Ges. 15: 2-10.

__________ 1898. Monographia Roccellearum. Biblioth. Bot. 45: 1-103.

__________ 1898a. Weiteres ueber die Flechtentribus der Roccellei. Ber. Deutsch. Bot. Ges. 16: 6-16.

__________ 1912. The lichens of the Swedish Antarctic Expedition. (In) Nordenskjold, Wissenschaftliche Ergebnisse der Schwedischen Suedpolar-Expedition 1901-1903, Bd. 4, Bot. No. 11, pp. 1-74. 2. Abt.

Degelius, Gunnar. 1954. The lichen genus Collema in Europe. Symb. Bot. Upsal. 13 (2): 1-499. 27 plates.

Du Rietz, G. Einar. 1926. Vorarbeiten zu einer "Synopsis Lichenum." Die Gattungen Alectoria, Oropogon, und Cornicularia. Ark. Bot. 20A (11):1-43. 2 plates.

__________ 1929. The discovery of an arctic element in the lichen flora of New Zealand and its plant geographical consequences. Austral. Assoc. Adv. Sci. for 1928, pp. 628-635. No species mentioned but good phytogeographical discussion.

Filson, Rex. B. 1965. Foliicolous lichen from south-east Gippsland. Victoria Naturalist 82: 68-69. Bacidia leucoloma.

__________ 1967. Supplementary descriptions for two Victorian desert lichens. Muelleria 1: 197-202.

__________ 1969. A review of the genera Teloschistes and Xanthoria in the lichen family Teloschistaceae in Australia. Muelleria 2: 65-115. 15 plates.

__________ 1970. Studies in Australian lichens. I. Victoria Naturalist 87: 324-327. Cladia and Hypogymnia.

Flinders, Matthew. 1801. Observations on the coasts of Van Diemen's Land, on Bass' Strait and its islands and on parts of the coasts of New South Wales. London.

__________ 1814. A voyage to Terra Australis, undertaken for the purpose of completing the discovery of that vast country, and prosecuted in the years 1801, 1802, and 1803, in His Majesty's Ship The Investigator, and subsequently in the . . . Porpoise and Cumberland Schooner. 2. vol. with atlas (folio). G. & W. Nicol, London. Lichens by Robert Brown.

Frey, Eduard. 1967. Die Lichenologischen Ergebnisse der Forschungsreisen des Dr. Hans Ulrich Stauffer (gestorb.) in Zentralafrika (Virunga-Vulkan 1954/55) und Suedafrika - Australien - Ozeanien - USA 1963-1964. Bot. Jahrb. Syst. 86: 209-255.

Fries, Elias. 1846-47. Plantae Preissianae sive Enumeratio Plantarum 2: 140-145. (Cheel 64).

Gyelnik, V. 1932. Peltigerae novae et rarae. Ann. Cryptog. Exot. 5: 39-40.

__________ 1932a. Was ist Solorina sorediifera Nyl? Loc. cit. 41-42.

Gyelnik, V. 1933. Lichenes varii novi criticique. Acta Fauna Fl. Universali, Ser. II, Bot. 1:3-10.

Hale, Mason E. 1960. A revision of the South American species of Parmelia determined by Lynge. Contr. U. S. Natl. Herb. 36: 1-41.

——— 1961. The typification of Parmelia perlata (Huds.) Ach. Brittonia 13: 361-367.

——— 1962. A new species of Parmelia from Asia: P. subcorallina. J. Jap. Bot. 37: 345-347.

——— 1963. The systematic position of Parmelia albata Wils. Bryologist 66: 72-74.

——— 1965. A monograph of Parmelia subgenus Amphigymnia. Contr. U. S. Natl. Herb. 36: 193-358.

——— 1965a. Studies on the Parmelia borreri group. Svensk Bot. Tidskr. 59: 37-48.

——— 1968. New Parmeliae from southeast Asia. J. Jap. Bot. 43: 324-327.

Hale, Mason E. & Syo Kurokawa. 1964. Studies on Parmelia subgenus Parmelia. Contr. U.S. Natl. Herb. 36(4): 121-191. Plates 1-9.

Hampe, E. 1843. Parmeliarum species tres novas. Linnaea 17: 120-123. P. lucaeana.

——— 1853, 1856. Plantae Muellerianae. Op. cit. 25: 709-712 (1853); 28: 216-217 (1856). (Cheel 20 with pagination incorrect).

Harmand, J. 1911, 1912. Lichens recueillis dans la Nouvelle Caledonie ou en Australie par le R. P. Pionnier. Bull. Soc. Sci. Nancy (3) 12: 124-143; 13: 37-63.

Hay, J. G. 1906. The visit of Mr. Charles Fraser, Colonial Botanist of New South Wales, with Captain Stirling, in H.M.S. "Success," to the Swan River in 1827, with his report on the botany, soil and capabilities of the locality. J. Western Australia Nat. Hist. Soc. 3: 16-35. No lichens, but a narrative of travels.

Hellbom, P. J. 1896. Lichenaea Neo-Zeelandia seu Lichenes Novae Zeelandiae a Sv. Berggren annis 1874-1875 collecti, additis ceteris speciebus indidem huc usque cognitis, breviter commemoratis. Bih. Kongl. Svenska Vetensk.-Akad. Handl. 21 (III, 13: 1-150. A secondary source only.

Henssen, Aino. 1963. Eine Revision der Flechtenfamilien Lichinaceae und Ephebaceae. Symb. Bot. Upsal. 18(1): 1-123. 31 plates.

——— 1965. A review of the genera of the Collemataceae with simple spores. (excluding Physma). Lichenologist 3: 29-41.

Hertel, H. 1969. Die Flechtengattung Trapelia Choisy. Herzogia 1: 111-130.

Hooker, W. J. 1842. On Cenomyce retipora. London J. Bot. 1: 292-294. tab. x, fig. 1-7. (Cheel 103).

Hue, A. M. 1890-92. Lichenes exotici a Professore W. Nylander descripti ... Ann. Sci. Nat. Bot. (3) 2: 211-322 (1890); 3: 33-192 (1891); 4: 103-156 (1892). (Cheel 133).

——— 1907. Heppiearum ultimae e familiae Collemacearum tribubus nonnullas species morphologice et anatomice elaboravit. Mém. Soc. Sci. Nat. Cherbourg 36: 1-44.

Hue. A. M. 1914. Plurimas lichenum species glaucogonidia continentes edisseruit. Bull. Soc. Bot. France 61: 333-340.

Imshaug, Henry A. & Irwin M. Brodo. 1966. Biosystematic studies on Lecanora pallida and some related lichens in the Americas. Nova Hedwigia 12: 1-59. tab. 1-10b.

Knight, C. 1871. Notes on Stictei in Kew Herbarium. J. Linn. Soc., Bot. 11: 243-246.

__________ 1882. Contribution to the lichenographia of New South Wales. Trans. Linn. Soc. London (2) 2: 37-51. (Cheel 7).

__________ 1884a. A description of a new Parmelia from Victoria. Proc. Roy. Soc. Queensland 1: 114. (Cheel 109).

__________ 1884b. Lichens. (In) Bailey, F.M. Contributions to the Queensland flora, Part II. Proc. Roy. Soc. Queensland 1: 84-92. (Cheel 70). Note: Part I, loc. cit. pp. 8-19 (Cheel 69) contains no lichen records.

__________ 1884c. Lichens. (In) Bailey, F.M. Contributions. . . Part III, loc. cit. pp. 151-153. (Cheel 71).

Koérber, G. W. 1862. Reliquiae Hochstetterianae. Abh. Schles. Ges. Vaterl. Cult., Abth. Naturwiss. 2: 30-34.

Krempelhuber, A. 1868. Exotische Flechten aus den Herbariums Wien. Verh. K. K. Zool.-Bot. Ges. Wien 18: 303-330. (Cheel 110).

__________ 1870. Lichenes. (In) Reise der oesterreichischen Fregatte Novara um die Erde in den Jahren 1857-9, Theil I, pp. 107-129. (Cheel 5).

__________ 1873. Beitrag zur Kenntnis der Lichen-Flora der Sued-See Inseln. J. Mus. Godeffroy 4: 93-100. (Cheel 111).

__________ 1880 (published 1881). Ein neuer Beitrag zur Flechtenflora Australiens. Verh. K.K. Zool.-Bot. Ges. Wien 30: 329-342. (Cheel 6) Note: The Catalogue sometimes erroneously cites this as 1860.

__________ 1880, 1881. Lichenes australiani e Baronis de Mueller collectionibus. (In) Meuller, F. von. Fragmenta phytogeographiae Australiae, vol. II, Suppl. 5, pp. 70-74 (1880); 115-118 (1881); 131-132 (1881).

Kurokawa, Syo. 1960. Anaptychiae (lichens) and their allies of Japan (5). J. Jap. Bot. 35: 353-358.

__________ 1962. A monograph of the genus Anaptychia. Nova Hedwigia, Beih. 6: 1-115. 9 plates.

__________ 1965. Revision of series Relicinae of the genus Parmelia in Japan and Taiwan. J. Jap. Bot. 40: 264-269.

__________ 1967. On the occurrence of diffractaic, physodalic, and psoromic acids in Parmeliae. Bull. Natl. Sci. Mus. 10: 369-375. Plate I-II.

__________ 1969. On the occurrence of norlobaridone in Parmeliae. J. Hattori Bot. Lab. 32: 205-216. Plate I-II.

__________ 1969a. A note on some rare lichens of Japan. J. Jap. Bot. 44: 225-229.

__________ 1969b. Lichenes Rariores et Critici Exsiccati, Fasc. II (Nos. 51-100). Scheda.

Kurokawa, Syo. 1971. Ditto, Fasc. III (Nos. 101-150). Scheda.

Kurokawa, Syo & John A. Elix. 1971. Two new Australian Parmeliae. J. Jap. Bot. 46: 113-116. Plate V.

Kurokawa, Syo, J.A. Elix, P. L. Watson & M.V. Sargent. 1971. Parmelia notata, a new lichen species producing two new depsidones. Loc. cit. pp. 33-36.

LaBillardiere, J. J. 1804. Novae Hollandiae Plantarum, Vol. 2, p.110.

Lamb, I. Mackenzie. 1947. A monograph of the lichen genus Placopsis Nyl. Lilloa 13: 151-288. Plates I-XVI.

__________ 1954. Studies in frutescent Lecideaceae (lichenized discomycetes). Rhodora 56: 105-129; 137-153.

__________ 1955. New lichens from northern Patagonia with notes on some related species. Farlowia 4: 423-471. 31 figs.

Laurer, F. 1827. Sieber'sche Lichenen. Linnaea 2: 38-46.

*Leighton, W.A. 1866. Notulae lichenologicae XI. On the examination and rearrangement of the Cladoniae, as tested by hydrate of potash. Ann. Mag. Nat. Hist. (3) 18: 405-420.

__________ 1867. Notulae Lichenologicae XII. On the Cladoniae in the Hookerian Herbarium at Kew. Op. cit. 19: 99-124.

Letrouit-Galinou, M.-A. 1957. Revision monographique du genre Laurera (Lichens, Trypetheliaceae). Rev. Bryol. Lichénol., n.s. 26: 207-264.

Lhotsky, John. 1843. Some data towards the botanical geography of New Holland. London J. Bot. 2: 135-141. Presents rough divisions of ecological units of the vegetation, mentioning no lichens.

Llano, George Albert. 1950. A monograph of the lichen family Umbilicariaceae in the Western Hemisphere. 281 pages. Office of Naval Research, Washington.

Magnusson, A. H. 1929. A monograph of the genus Acarospora. Kongl. Svenska Vetensk.-Akad. Handl. (3) 7 (4): 1-400.

__________ 1940. Studies in species of Pseudocyphellaria. The Crocata-group. Acta Horti Gothob. 14: 1-36.

Maiden, J. H. 1898. Notes of a trip to Mt. Seaview, Upper Hastings River. Proc. Linn. Soc. New South Wales 23: 24-25. (Cheel 15).

__________ 1899. A second contribution to the flora of Mt. Kosciusko. Agric. Gaz. New South Wales 10: 1001-1042. (Cheel 141). Merely mentions "a green lichen and a black foliaceous one."

*__________ 1902-03. The Sydney Botanic Garden. Biographical notes concerning the officers in charge. Public Service Journal (Sydney) (with supplements).

__________ 1908. Records of Victorian botanists. Victoria Naturalist 25: 101-117. A good historical summation, no species mentioned.

*__________ 1908a. A century of botanical endeavor in South Australia (address by the President, Section D, Biology). Rep. Austral. Assoc. Adv. Sci. 11: 158-199. (for 1907). Cited by Maiden as "Notes on South Australian Botanists." Many notes and references given.

Maiden, J. H. 1908b. Records of Australian botanists (a) General, (b) New South Wales. J. Proc. Roy. Soc. New South Wales 42: 60-132. 5 plates.

*__________ 1909. Records of Australian botanists who have dealt with the flora of western Australia. J. West Austral. Nat. Hist. Soc. 6: 28-34.

*__________ 1910. Records of Queensland botanists. Rep. Austral. & N. Z. Assoc. Adv. Sci. 12: 373-383. 2 plates.

*__________ 1910a. Records of the earlier French botanists as regards Australian plants. J. Proc. Roy. Soc. New South Wales 44: 123-155. 11 plates.

__________ 1912. Records of Australian botanists (First Supplement). Rep. Austral. Assoc. Adv. Sci. for 1911, pp. 224-243.

Martin, Wm. 1962. Notes on some New Zealand species of _Cladonia_ with descriptions fo two new species and one new form. Trans. Roy. Soc. New Zealand, Bot. 2: 39-44.

__________ 1962a. The lichen genus _Cladia_. _Op_. _cit_. 3: 7-12.

Montagne, C. & M. J. Berkeley. 1846. On _Thysanothecium_, a new genus of lichen collected by Drummond in Swan River. London J. Bot. 5: 257. (Cheel 62).

Motyka, J. 1936-38. Lichenum generis _Usnea_ studium monographium. Pars. systematica, Vol. I, II.

__________ 1960. Zmiennosc _Thamnolia vermicularis_ (Sw.) Schaer. De variabilitate _Thamnoliae vermicularis_ (Sw.) Schaer. Fragm. Florist. Geobot. 6: 627-635.

*Mueller, Baron Ferdinand von. 1854. Report, Botanic Garden, Melbourne, Legislative Council, pp. 18-19. (Cheel 21).

*__________ 1858. Report, Botanic Garden, Melbourne, Legislative Assembly, p. 13. (Cheel 22).

__________ 1858. Report on the plants collected during Mr. Babbage's expedition into the Northwestern Interior of South Australia in 1858. Victoria. 21 pages. One lichen mentioned.

__________ 1877. List of the plants obtained during Mr. Giles' travels in Australia in 1875 and 1876. J. Bot. 15: 344-349. (Cheel 105).

__________ 1880. See Krempelhuber.

__________ 1881. Lichens. (In) Fragmenta Phytogeographiae Australiae, Melbourne. 11 (addition to Suppl. 5): 115-118, 131-132. Rewriting of other literature reports only, a secondary source.

__________ 1881, 1883. Census of the genera of plants hitherto known as indigenous to Australia. J. Proc. Roy. Soc. New South Wales 15: 185-300 (lichens pp. 249-251) (1881); and suppl., 17: 187-189 (1883). (Cheel 104). Only genera are mentioned.

__________ 1887a. List of Australian lichens, indicative of additional species or of unrecorded localities, from Dr. J. Mueller's elucidations. Victoria Naturalist 4: 88-95. (Cheel 9). A secondary source, the taxa listed being copied from other publications or from unpublished correspondence.

Mueller-Arg., J. 1878. Lichenologische Beitraege VII. Flora 61: 481-492.

Mueller-Arg., J. 1879. Ditto, IX. _Op_. _cit_. 62: 289-298.

__________ 1881a. Ditto, XII. _Op_. _cit_. 64: 81-88; 100-112.

__________ 1881b. Ditto, XIV. _Op_. _cit_. 64: 513-527.

__________ 1882a. Ditto, XV. _Op_. _cit_. 65: 291-306; 316-322; 326-337; 381-386; 397-402.

__________ 1882b. Ditto, XVI. _Op_. _cit_. 65: 483-490; 499-505; 515-519.

__________ 1883a. Ditto, SVII. _Op_. _cit_. 66: 17-25; 45-48; 75-80.

__________ 1883b. Ditto, XVIII. _Op_. _cit_. 66: 243-249; 271-274; 286-290; 304-306; 317-322; 330-338; 344-354.

__________ 1884a. Ditto, XIX. _Op_. _cit_. 67: 268-274; 283-289; 299-306; 349-354; 396-402; 460-468.

__________ 1884b. Ditto, XX. _Op_. _cit_. 67: 613-621.

__________ 1885. Ditto, XXII. _Op_. _cit_. 68: 499-502; 503-518; 528-533.

__________ 1886. Ditto, XXIV. _Op_. _cit_. 69: 252-258; 286-290; 307-318.

__________ 1887a. Ditto, XXV. _Op_. _cit_. 70: 56-64; 74-80.

__________ 1887b. Ditto, XXVI. _Loc_. _cit_. pp. 268-273; 283-288; 316-322; 336-338; 396-402; 423-429.

__________ 1887c. Revisio lichenum Australiensium Krempelhuberi. _Op_. _cit_. 70: 113-118. (Cheel 119). These are corrections of determinations appearing earlier in Krempelhuber's papers.

__________ 1888a. Lichenologische Beitraege XXVII. _Op_. _cit_. 71: 17-25; 44-48.

__________ 1888b. Ditto XXVIII. _Loc_. _cit_. pp. 129-142.

__________ 1888c. Ditto XXIX. _Loc_. _cit_. pp. 195-208.

__________ 1888d. Ditto XXX. _Loc_. _cit_. pp. 528-552.

__________ 1889. Ditto XXXII. _Op_. _cit_. 72: 505-508.

__________ 1890a. Lichenes epiphylli novi. Geneva. 22 pages. (Cheel 120).

__________ 1890b. Lichenologische Beitraege XXXIII. Flora 73: 187-202.

__________ 1890c. Lichenes Africae tropico-orientalis. _Loc_. _cit_. pp. 334-347. (Cheel 121). There are, however, no Australian entries.

__________ 1891a. Lichenes Brisbanenses, a cl. F. M. Bailey, prope Brisbane Queensland in Australiae orientale lecti. Nuovo Giorn. Bot. Ital. 23: 385-404. (Cheel 123).

__________ 1891b. Lichenes Bellendenici a cl. F. M. Bailey, Govt. Bot. ad Bellenden-Ker Australiae orientalis lecti et sub numeris citatis missi. Hedwigia 30: 47-56. (Cheel 122).

__________ 1891c. Lichenologische Beitraege XXXIV. Flora 74: 107-118.

__________ 1891d. Ditto XXXV. _Loc_. _cit_. pp. 371-382.

__________ 1892a. Lichenes exotici Herbarii Vindobonensis. I. Lichenes in Australia et in ejus vicinitate lecti. Ann. K. K. Naturhist. Hofmus. 7: 302-305. (Cheel 124).

__________ 1892b. Lichenes Australiae occidentalis a cl. Helms recenter lecti et a celeb. Bar. Ferd. v. Mueller comm. Proc. Roy. Soc. South Australia 16: 142-149 (also reprinted in Hedwigia 5: 191-198). (Cheel 127).

__________ 1893a. Lichenes Wilsonianae s. Lichenes a cl. Rev. F. R. M. Wilson in Australia Prov. Victoria lecti. Bull. Herb. Boiss. 1: 33-65. (Cheel 126). Zahlbruckner in the Catalogus consistently gives the date as 1892.

__________ 1893b. Lichenes Exotici, II. Hedwigia 32: 120-136. (Cheel 125). Lichenes Exotici I has no Australian entries.

__________ 1895a. Ditto, III. Op. cit. 34: 27-38. (Cheel 125 in part).

__________ 1895b. Pyrenocarpae Queenslandiae. Rep. Austral. Assoc. Adv. Sci. pp. 449-466. (Cheel 129).

__________ 1895c. Sertum Australiense s. species novae Australienses Thelotremarum, Graphidearum, et Pyrenocarpearum. Bull. Herb. Boiss. 3: 313-327. (Cheel 130).

__________ 1895d. Lecanorae et Lecideae Australienses novae. Loc. cit. pp. 632-642. (Cheel 128).

__________ 1896. Analecta Australiensia. Op. cit. 4:87-96. (Cheel 13).

__________ 1898. Lichenes Australienses. Op. cit. 6:78-80. (Cheel 14). Posthumously prepared by F. R. S. Wilson.

Murray, J. 1960a. Studies of New Zealand lichens. Part I, the Coniocarpineae. Trans. Roy. Soc. New Zealand 88: 177-195.

__________ 1960b. Ditto, Part II, the Teloschistaceae. Loc. cit. pp. 197-210.

__________ 1960c. Ditto, Part III, the family Peltigeraceae. Loc. cit. pp. 381-399.

Nylander, W. 1857 (issued 1858). Enumération générale des lichens, avec l'indication sommaire de leur distribution géographique. Mem. Soc. Scienc. Imp. Acad. Cherbourg 5: 84-146; suppl. pp. 332-339.

__________ 1864. Circa G. W. Koerberi reliquas Hochstetterianas. Flora 47: 266-270.

__________ 1868-1870. Conspectus synopticum Sticteorum. Bull. Soc. Linn. Normand. (2) 2: 47 (1868); 4: 167 (1870). (Cheel 107). There are no pertinent entries on Australian species.

__________ 1870. Recognitio monographica Ramalinarum. Bull. Soc. Linn. Normand. (2) 4: 101-180.

__________ 1885. Parmeliae exoticae novae. Flora 68: 605-615 (Cheel 108).

__________ 1886. Lichenes nonnulli Australienses. Op. cit. 69: 323-328. (Cheel 106, gives date incorrectly as 1866).

Persoon, C. H. 1826. Lichens. (In) Gaudichaud, C. Voyage autour du monde exécuté sur les corvettes de S. M. l'Uranie et la Physicienne, L. Freycinet, Capt. de vaisseau. Part 4 (Botanique), i-viii + 522 pp. Lichens, pp. 187-215. (Cheel 2).

Raesanen, Veli, 1948 (1949). Lichenes Novi. V. Arch. Soc. Zool. Bot. Fenn. "Vanamo" 3: 178-188.

Santesson, Rolf. 1942. The South American Menegazziae. Ark. Bot. 30A (11): 1-35. Plates I-II.

__________ 1952. Foliicolous lichens I. A revision of the taxonomy of the obligately foliicolous, lichenized fungi. Symb. Bot. Upsal. 12 (1): 1-590.

Shirley, J. 1888a. Additions to the lichen flora of Queensland. Proc. Roy. Soc. Queensland 5: 7-10, 72-75. (Cheel 78, 78a).

__________ 1888b. Field naturalists' excursion to Caboolture. Loc. cit. pp. 137-144, lichens p. 142. (Cheel 79).

__________ 1889a. The lichen flora of Queensland. Op. cit. 5: 80-110 (1888); 6: 3-55, 129-145, 165-218 (1889). (Cheel 84).

__________ 1889b. Additions to the lichen flora of Queensland. Loc. cit. pp. 115-116. (Cheel 83).

__________ 1889c. (In) Simmonds, J. H. Field Naturalists' excursion [near Powder Magazine at Eagle Farm, Queensland]. Loc. cit. pp. 64-65. (Cheel 80). Lichens det. by Shirley, p. 65.

__________ 1889d. (In) Simmonds, J. H. Excursion of Field Naturalists to Brookfield. Loc. cit. pp. 65-70. (Cheel 81). Lichens det. by Shirley, pp. 69-70.

__________ 1889e. New or rare lichens (exhibited). Loc. cit. p. 80. (Cheel 82).

*__________ 1890. The lichen flora of Queeensland. (Cheel 85). "This is a separate work giving descriptions of 485 species. Included in this are those published in the Proc. Roy. Soc. Queensland." (Cheel 1906).

__________ 1891a. Lichenes (In) Bailey, F. M. Contributions to the Queensland flora. Queensland Dept. Agr. Bull. 9: 20-32 (Bailey's Bot. Bull. No. 3). (Cheel 87 in part).

__________ 1891b. Lichenes (In) Bailey, F. M., ditto. Op. cit. 13: 23-25 (Bot. Bull. No. 4). (Cheel 87 in part).

__________ 1892a. Lichenes (In) Bailey, F. M., ditto. Op. cit. 18: 31-35 (Bot. Bull. No. 5). (Cheel 88).

__________ 1892b. Lichens from Warwick and neighborhood. Proc. Roy. Soc. Queensland 8: 132-135. (Cheel 86, giving date as 1891).

__________ 1893a. Lichenes. (In) Bailey, F. M. Contributions to the Queensland flora. Queensland Dept. Agr. Bull. :91-106 (Bot. Bull. No. 8). (Cheel 89).

__________ 1893b. The geographical distribution of Queensland lichens. Trans. Austral. Assoc. Adv. Sci. 5: 410-413. (Cheel 92). No lichen records given.

__________ 1894. On some Victorian lichens. Proc. Roy. Soc. Queensland (2) 5: 54-58. (Cheel 35).

__________ 1895. Lichenes. (In) Bailey, F. M. Contributions to the Queensland flora. Queensl. Dept. Agr. Bull. :28-35 (Bot. Bull. No. 10). (Cheel 90).

Shirley, J. 1896. Lichens. (In) Bailey, F. M. ditto. Op. cit. :19-31 (Bot. Bull. No. 13). (Cheel 91).

Smith, G. G. 1962. The flora of granite rocks of the Porongurup Range, Southwestern Australia. J. Roy. Soc. Western Australia 45: 18-23. Cited by Henssen (1963) as describing the type locality of Porocyphus lichinelloides.

*Stirton, J. 1873, 1875, 1876. Lichens British and foreign. Trans. Field-Nat. Soc. Glasgow 4: 85-95 (1875), other pages not given. (Cheel 113).

__________ 1877-78. On certain lichens belonging to the genus Parmelia. Scott. Naturalist (Perth) 4: 200-203; 252-254; 298-299. (Cheel 114).

*__________ 1878. (Title?). Trans. Field-Nat. Soc. Glasgow, Jan., April, July. (Cheel 72).

__________ 1880. Lichens collected by Mr. Hugh Paton, during a tour through Southern Australia (Gippsland), communicated by J. E. Tenison-Woods. Trans. Proc. Roy. Soc. Victoria 17: 10-13. (Cheel 115).

__________ 1881a. (issued 1882). A new genus of lichens. Op. cit. 18: 1-2. (Cheel 74). Trichocladia (=Heterodea).

__________ 1881b. Additions to the lichen flora of Queensland. Op. cit. 17: 66-78. (Cheel 73).

__________ 1881c. On the genus Usnea, and another (Eumitria) allied to it. Scott. Naturalist (Perth) 6: 99-109. (Cheel 114).

__________ 1882. Notes on the genus Usnea, with descriptions of new species. Loc. cit. pp. 292-297; op. cit. (n.s.) 1: 74-79 (1883). (Cheel 114 in part).

__________ 1888. Lichens. Op. cit. pp. 307-309.

__________ 1897 (issued 1898). On new Australian and New Zealand lichens. Trans. Proc. New Zealand Inst. 30: 382-393. (Cheel 75, 117 in part).

__________ 1899a (issued 1900). On new lichens from Australia and New Zealand. Op. cit. 32: 70-82. (Cheel 117 in part).

__________ 1899b. Lichens. (In) Bailey, F. M. Contributions to the Queensland flora. Queensland Agr. J. 5: 37-41; 484-488. (Cheel 118).

Stizenberger, Ernst. 1895. Die Gruebflechten (Stictei) und ihre geographische Verbreitung. Flora 81: 88-150. (Cheel 132).

Tate, R. 1881 (issued 1882). A list of the charas, mosses, liverworts, lichens, fungi and algae of extra-tropical South Australia. Trans. Proc. Roy. Soc. South Australia 4: 5-24 (Lichens, p. 9). Names extracted from Krempelhuber's list in Mueller's Fragmenta, probably a wholly secondary source.

__________ 1882. Additions to the flora of South Australia. Op. cit. 5: 82-93 (lichens, p. 92). (Cheel 58).

__________ 1887. Additional lichens and fungi of South Australia collected by J. G. O. Tepper, F. L. S., 1880 to 1885. Op. cit. 9: 215-216. (Cheel 60, entitled "Lichens from St. Vincent's Gulf, York Peninsula, Clarendon and Kangaroo Island").

Taylor, T. 1844. Descriptions of new mosses and lichens from the Australian colonies. Phytologist 1: 1093-1096.

__________ 1847. Lichens, principally from the herbarium of Sir William Hooker. London J. Bot. 6: 148-193. (Cheel 63). Swan River lichens collected by Drummond.

Tepper, J. G. O. 1883. Botanical notes (three lichen species from Clarendon). Proc. Roy. Soc. South Australia 6:66. (Cheel 59).

Turner, F. 1905. (A list of 37 species of lichens, principally from New England, N. S. W.). Proc. Linn. Soc. New South Wales 30: 259, 308-311. (Cheel 143).

Vainio, E. 1887-1897. Monographia Cladoniarum Universalis. Acta Soc. Fauna Fl. Fenn. 4: 1-509 (1887); 10: 1-499 (1894); 14: 1-268 (1897). (Cheel 187).

__________ 1888. De subgenere Cladinae. Medd. Soc. Fauna Fl. Fenn. 14: 31-32. (Cheel 138). No reference made to Australia.

__________ 1898. Clathrinae herbariae Muelleri. Bull. Herb. Boiss. 6: 752. (Cheel 139).

__________ 1900. Reactiones lichenum a J. Muellero Argoviensi descriptorum. Mem. Herb. Boiss. 5: 1-17.

Verseghy, Klara. 1962. Die Cattung Ochrolechia. Nova Hedwigia Beih. 1: 1-146. 12 plates.

Watts, Rev. W. W. 1903. A list of 27 lichens, determined by Dr. Bouly de Lesdain of Dunkerque. Proc. Linn. Soc. New South Wales 28: 498-499. (Cheel 19).

Weber, William A. 1968. A taxonomic revision of Acarospora, subgenus Xanthothallia. Lichenologist 4: 16-31.

__________ 1969. Lichenes Exsiccati distributed by the University of Colorado, Fasc. 6-7, No. 201-280. Scheda.

__________ 1971. Ditto, Fasc. 8-9, No. 281-360. Scheda.

Wetmore, Clifford M. 1963. Catalogue of the lichens of Tasmania. Rev. Bryol. Lichenol. 32: 223-264.

Willey, Henry. 1890. A synopsis of the genus Arthonia. Published by the author, New Bedford, Mass. 62 pp.

Willis, James H. 1953. The Archipelago of the Recherche (Western Australia), Part 3a - Land flora. Austral. Geogr. Soc. Rep. (Melbourne) 1: 1-35 (lichens pp. 31-32, 34-35).

__________ 1956. A new genus of alpine lichens. Victoria Naturalist 73: 125-128. Bibbya muelleri, based on Siphula muelleri Wils.

__________ 1959a. Reduction of the lichen genus Bibbya J. H. Willis. Muelleria 1: 91-92.

__________ 1959b. Notes on the vegetation of Eucla District, Western Australia. Loc. cit. pp. 92-96.

Wilson, Rev. F. R. M. 1887. Notes on a few Victorian lichens. Victoria Naturalist 4: 83-87. (Cheel 26).

__________ 1888. Descriptions of two new lichens, and a list of additional lichens new to Victoria. Op. cit. 5: 29-31. (Cheel 27).

Wilson, Rev. F. R. M. 1889a. An additional list of lichens new to Victoria, with a description of 41 Victorian lichens new to science. Op. cit. 6: 60-61. (Cheel 28). Nomina nuda.

__________ 1889aa. A. description of forty-one Victorian lichens new to science. Loc. cit. pp. 61-69. (Cheel 29). Zahlbruckner evidently did not see this validation of the nomina nuda given in the previous paper. Many of these are validly published herein.

__________ 1889b. Notes on lichens in New South Wales. Proc. Roy. Soc. Queensland 6: 85-93. (Cheel 10 and 11).

__________ 1889c. A hunt for lichens in East Gippsland. Victoria Naturalist 6: 57-59. (Cheel 30).

__________ 1890. Australian lichenology. Rep. of second meeting of Austral. Assoc. Adv. Sci. 3: 549-553. Essay on history and collectors.

1890a. Lichens from the Victorian Alps. Victoria Naturalist 6: 178-180. (Cheel 31).

__________ 1890b. Lichens from Western Australia. Loc. cit. p. 180. (Cheel 65).

__________ 1891a. Notes on a remarkable lichen growth in connection with a new species of Sticta with description of both. Proc. Roy. Soc. Queensland 7: 8-11. (Cheel 32).

__________ 1891b. On lichens collected in Victoria, Australia (communicated by Mr. Carruthers). J. Linn. Soc. Bot. 28: 353-374. 1 plate (Cheel 33).

__________ 1891c. Lichenes. (In) Bailey, F. M. Contributions to the Queensland flora. Queensland Dept. Agr. Bull. 7: 28-33 (Bailey's Bot. Bull. No. 2). (Cheel 93, entitled "A lsit of Queensland lichens new to science.").

__________ 1902. The lichens of Victoria, Part I. Proc. Roy. Soc. Victoria (2) 5: 144-177. (Cheel 34).

Woolls, W. 1867. Contribution to the flora of Australia. Sydney, x + 225 pp. (Cheel 4). Lichens pp. 163-173.

Zahlbruckner, Alexander. 1896. Lichenes Mooreani. Ann. K. K. Naturhist. Mus. 11: 187-196. (Cheel 12).

__________ 1899. VII. Flechten. B. Verzeichniss der neuen Gattungen, Arten und Varietaeten. Bot. Jahresber. 27: 439-449. A secondary reference only.

__________ 1900. Schedae ad "Kryptogamas Exsiccatas" editae a Museo Palatino Vindobonensi, Cent. V-VI. Ann. K. K. Naturhist. Mus. 15: 169-215.

*__________ 1902-1904. Lichenes rariores exsiccatae. Schedae.

__________ 1903. Schedae ad "Kryptogamas Exsiccatas" editae a Museo Palatino Vindobonensi, Cent. IX. Ann. K. K. Naturhist. Mus. 18: 349-375.

__________ 1904a. Neue Flechten. Ann. Mycol. 2: 267-270. (Cheel 136).

__________ 1904b. Schedae ad "Kryptogamas Exsiccatas" editae a Museo Palatino Vindobonensi, Cent. X-XI. Ann. K. K. Naturhist. Mus. 19: 379-427.

Zahlbruckner, Alexander. 1905. Ditto, Cent. XII-XIII. *Op. cit.* 20: 311-358.

__________ 1910. Ditto, Cent. XVIII. *Op. cit.* 24: 269-292.